MODERN PHYSICS AND ITS PHILOSOPHY

SYNTHESE LIBRARY

MONOGRAPHS ON EPISTEMOLOGY,

LOGIC, METHODOLOGY, PHILOSOPHY OF SCIENCE,

SOCIOLOGY OF SCIENCE AND OF KNOWLEDGE,

AND ON THE MATHEMATICAL METHODS OF

SOCIAL AND BEHAVIORAL SCIENCES

M. STRAUSS

MODERN PHYSICS
AND ITS PHILOSOPHY

Selected Papers in the
Logic, History, and Philosophy of Science

D. REIDEL PUBLISHING COMPANY / DORDRECHT-HOLLAND

Library of Congress Catalog Card Number 71-183369

ISBN-13:978-94-010-2895-0   e-ISBN-13:978-94-010-2893-6
DOI: 10.1007/978-94-010-2893-6

PREFACE

In selecting the papers for this volume I have excluded all physics papers proper. I have further omitted all book reviews. Instead, I have included two papers not published previously; they are marked by an asterisk (∗) in the table of contents.

Since many of the papers were occasioned by Symposia or similar gatherings their chronological order is rather accidental. Hence I have tried to group the papers thematically into four parts. Within each part the order of sequence is from the more general to the more special, or from a more popular to a more technical treatment. The same principle has been applied to the sequential order of the parts. The foundational papers on quantum mechanics have been arranged in a somewhat different manner. Chapters XVI–XIX are concerned with the logic of complementarity while in Chapters XX–XXII a more radical reconceptualization is carried out.

Two of the older papers (Chapters VI and VIII) have been revised to bring them more into line with present terminology. Other papers have been corrected by additions and omissions. Additions are marked by square brackets [ ], while double square brackets [[ ]] signify omissions or parts to be omitted. Hence $[[A]]$ $[B]$ means that 'A' should be replaced by 'B'.

The heading of one paper (Chapter XX) has been changed to make it more descriptive.

My thanks are due to the Editors, particularly to Professor J. Hintikka, and to the Publishers for their understanding and patience in preparing this publication.

THE AUTHOR

ACKNOWLEDGEMENTS

Permission to use copyrighted material for reprinting or translation has been granted by
The Rector of the Humboldt University at Berlin
VEB Deutscher Verlag der Wissenschaften, Berlin
Akademie-Verlag, Berlin
B. G. Teubner Verlagsgesellschaft, Leipzig
and is gratefully acknowledged.

The Publisher   *The Author*

A. REIDEL   M. STRAUSS

CONTENTS

PART A

HISTORY OF PHYSICS

CHAPTER I

CONTRADICTION AND UPLATION IN
THE EVOLUTION OF PHYSICS*

Traditional historiography of science creates the impression that the evolution of physics is determined above all by the ingenious ideas of a few great physicists, an impression enhanced by various 'scientific auto-biographies'. In contrast to this, the present paper subscribes to the view that the evolution of physics is subject to objective laws and hence fore-seeable, at least in principle, as to its general trend, though not, of course, in its details. Knowledge of evolutionary trends or laws not only facili-tates orientation in topical controversies but may serve as an important tool in planning research.

The general laws of evolution also apply to the evolution of physics, as a study of its history reveals. According to the basic law of evolution the driving force of the latter is the unfolding and overcoming of con-flicts or contradictions.[1] In physics, we have to do with the following kinds of contradictions:

1. contradictions between theory and facts (*external contradictions*);

2a. contradictions between different parts of accepted theory or, as we usually say, between different theories accepted at the same time (e.g., between thermodynamics and Maxwell's electrodynamics: *ultra-violet catastrophe* in the spectral distribution problem for black-body radiation)[2];

2b. contradictions turning up in the application of a theory to certain problems (e.g., the *cosmological paradox* in the Newtonian theory of gravitation, *Gibbs' paradox* in the classical statistical theory of diffusion, the *exploding electron* [in the Maxwell-Lorentz theory], *Klein's paradox* in the original version of Dirac's theory of the electron, *divergencies* in quantum field theory).

The contradictions in types (2a) and (2b) may be called *internal con-tradictions.*

While the role played by the external contradictions in the evolution of physics is generally known, the internal contradictions are often re-garded as minor flaws to be removed by slightly modifying the given theory. In contrast to this view, the history of physics shows that most of these internal contradictions can only be overcome by entirely novel theories; they should therefore be regarded as germs of, or [in terms of heuristics] as guides to, a new fundamental theory. A similar remark applies of course to the external contradictions.

Before the new theory is found, provisional solutions are usually offered in the form of ad hoc hypotheses [or methods]. Typical examples are Lorentz's contraction hypothesis, the Bohr-Sommerfeld quantum conditions, the renormalization in the quantum theory of fields. Such methods of dealing with the (external or internal) contradictions *within* the frame of the old theory can be considered as *partial anticipation* of a new theory. The use within quantum mechanics of Bose-Einstein statistics [for integer spin particles] and of Fermi-Dirac statistics [for half-integer spin particles] has also turned out to be a partial anticipation of a deeper theory, viz., the quantum theory of fields.

A new theory that resolves the old contradictions is always a generalization or rather *uplation* [dialectic negation] of the old theory: the old theory is dismissed [as a fundamental theory] and replaced by the new one, but some essential features of the old theory are taken over into the new one. – The term 'generalization' [either] refers to the mathematical formalism [[and]] [or] is meant to express that the range of validity of the new theory is larger than that of the old one. It does not, however, express clearly that from the standpoint of the new theory the old one is *conceptually degenerate* in the sense that different concepts have the same extension. The removal of this degeneracy [i.e. the splitting up of degenerate concepts] is an essential aspect of theory uplation.

In the evolution of physics the general law of evolution – unfolding and overcoming of contradictions – largely coincides with the trend towards unification of fundamental theories. Thus Einstein's Special Theory, in removing the contradiction between mechanics and electrodynamics, has established a general kinematic basis for all parts of physics [excluding gravitation] while quantum theory has removed the contradictions between thermodynamics and the other parts of physics. As well known, quantum theory originated from Planck's endeavour to unify thermodynamics and [Maxwell's] theory of the electromagnetic field. Thus, the struggle for a unified world picture is vindicated by the objective development of physics; yet it proves to be progressive and successful only in so far as it coincides with the objective trend, i.e., with the historical requirements and possibilities given at that time. If the struggle for a unified world picture goes beyond these requirements and possibilities we usually get abortive speculations such as the phlogiston theory of heat, the mechanical theories of the 'light aether', or the various 'unified field theories'. Sometimes the justified endeavour for unification turns into the metaphysical attempt at guessing the 'world formula' – presumably meaning the formula according to which God has created the world. Such tendencies, characteristic of objective idealism, can be found in Eddington's work and in the late work of Einstein; they seem to have been the deeper reason for Einstein's negative attitude towards quantum mechanics.

Most of the physicists living today would agree that cognition of Nature is a process with no end in sight, with periods of revolution alternating with periods of consolidation. The driving force is usually supplied by *external* contradictions – Einstein's General Theory being an exception rather than the rule. Yet for finding the new theory the *internal* contradictions [or even the 'weak spots'] of the old theory acquire heuristic importance in the same increasing measure in which the basic concepts of physical theory become more and more remote from the concepts of everyday physics. With increasing remoteness, *scientific abstraction* gains increasing importance in comparison to *induction*, as is generally recognized; but it is equally true that the existing theory, too, [with its internal contradictions] gains increasing heuristic importance in comparison to new discoveries [[...]]. As the latter rarely indicate the direction in which the given theory is to be generalized [or rather: uplated] in a *unique* way, and as the internal contradictions of the given theory are but an expression of its provisional character, the task of overcoming these contradictions largely coincides with the task of accounting for the new experimental facts by a novel theory. This is precisely what has happened both in the transition to quantum mechanics and in the transition to Einstein's Special Theory, though it was realized only afterwards what internal contradictions were removed in each case.

Today, theoretical physics is confronted with the task of uplating the existing quantum theory of fields by a theory of fundamental particles. The peculiarity of the situation appears to lie in the following. First, external contradictions are almost absent while the internal contradictions are rather well-known. Secondly, the new theory is expected to solve three rather different problems:

1. *system of fundamental particles:* the new theory is required to yield a classification scheme, comparable to the Periodic System of chemical elements, containing all known fundamental particles and perhaps predicting other such particles and their properties;

2. *coupling constants:* the new theory is required to yield the numerical values of the coupling constants [of present theory] which determine the magnitude of the interaction between the various particle fields and hence the mean life time of particles pertaining to these fields;

3. *mass spectrum:* the new theory is required to yield correct mass ratios for all fundamental particles.

Clearly, only the first of these problems may possibly be solved by induction and scientific abstraction on the basis of existing experimental data and without the heuristic clues furnished by present quantum field theory and its internal contradictions. However, just as the Periodic System of chemical elements called for a theoretical foundation that was only furnished by quantum mechanics, the mere establishment of a system of fundamental particles by inductive methods would hardly signify a significant advance in fundamental theory.[3]

It is in line with this that the new theoretical approaches aim at a generalization [or uplation] of present theory. Their dicussion would transcend the bonds of the present contribution. Howevcr, I will not conceal my conviction that a non-local generalization of the present field concept appears to be more in line with the general trend of the evolution of physics than a non-linear generalization of the field equations. Presumably, neither the one nor the other generalization will solve all problems listed above; indeed, this is hardly to be expected since these problems are of very different character and may perhaps be solved only by a sequence of successive generalizations. In particular, the novel laws of conservation [or quasi-conservation] [of quantities referring to internal degrees of freedom], while of fundamental importance for the systematization of fundamental particles, do not appear to have a direct import on the problem of the mass spectrum [as the latter is no doubt a dynamical rather than a kinematical problem].[4]

Whatever the solution of these problems may be, there is no doubt that it will be even more remote from the old mechanical materialism than present theory and that it will present a further vindication and evolution of dialectical materialism.

NOTES

* Translated from 'Widerspruch und Aufhebung in der Entwicklung der Physik' in *Naturwissenschaft und Philosophie* (ed. by G. Harig and F. Schleifstein), Berlin 1960.
[1] The term 'contradiction' usually has a more specific meaning than the German 'Widerspruch' which often means 'conflict'. The reader is asked to accept 'contradiction' as being equivalent to 'Widerspruch'.
[2] [It is sometimes suggested that the *ultraviolet catastrophe* – a consequence of Rayleigh-Jeans' spectral distribution formula – merely reveals a contradiction between Maxwell's theory and the equipartition theorem of classical statistical mechanics or even merely the inapplicability of the latter to a system with an infinite number of degrees of freedom. While it is true that the equipartition theorem and the Rayleigh-Jeans formula stand or fall together, the point to be made here is that the Rayleigh-Jeans formula follows already from *dimensional analysis* if the Maxwell theory is accepted.]
[3] [In the meantime the problem of systematic classification has been largely solved for baryons by a combination of induction, scientific abstraction, and group theory (SU_3 and other groups), finally leading to the prediction of 'quarks'. Whether this scheme will be vindicated by the final theory is anybody's guess.]
[4] [Meanwhile, certain mass relations involving more than two masses could be theoretically established on the basis of the group theoretical classification schemes; a complete explanation of the mass spectrum has not yet been found and can hardly be expected except as a consequence of a full-blown theory. However, contrary to what is surmised in the text, the masses of the fundamental particles should be expected to be functions of the internal degrees of freedom (and possibly other parameters), and this seems to be borne out by the results obtained so far.]

CHAPTER II

EVOLUTIONARY LAWS
AND PERSPECTIVES FOR PHYSICS*

The following considerations are a first attempt at formulating [internal] *laws for the historical evolution of physics* – a task as yet unsolved by historiography [of science]. The practical value of this attempt resides in the fact that [the knowledge of] such evolutionary laws provides a scientific basis for *prognostics* and *research strategy* that cannot be obtained in any other way.

Research in theoretical physics must of course rely in the first place on the experimental facts. These facts do not, however, provide more than circumstantial evidence, as Einstein put it. How this evidence is to be judged depends on *superior principles* used in the [theoretical] interpretation of it.

Of course, one can always use the principles of *dialectical materialism* as superior guides, as their range of validity is almost unlimited. However, range of validity and semantic content stand, for logical reasons, in a reciprocal relation. Hence, the laws of dialectical materialism are *too general* for the purposes of prognostics and research strategy.

Recognizing this state of affairs, philosophers [of science] are turning more and more towards the *methodology of science* – a trend noticeable also in this Symposium. I welcome this as a step in the right direction. It is no doubt useful to study the methods of gaining knowledge and of constructing theories. How much, however, can be gained by such studies for a scientific foundation of prognostics and research strategy? Are we sure that the methods used in the past will be successful in the future? The heuristic considerations of Schroedinger and Einstein which led to [one form of] quantum theory and the present chronogeometrical theory of gravitation, respectively, do not form part of these theories, as we know today. If such heuristic considerations are to be excluded from the methodology of science they have to be replaced by something better. If they are to be retained, this would mean we would have to rely even more on the good luck and genius of [individual] scientists. Furthermore, the methods of gaining knowledge and constructing theories are subjected to historical change; e.g., induction is loosing ground to mathematical construction. Methodology thus turns into a *historiography of methods*, and the history of these methods is determined, in the last analysis, by the evolutionary laws of physics. The latter thus provide the *key* to the methodology of physics: the reverse would be subjective idealism. This is not all: knowledge of the laws operative in the development of a science provides a scientific basis for prognostics and research strategy, and thus yields a *new method for gaining knowledge*. For these reasons, rationalization of research, now a general aim and even a social necessity, will be better served when we rely on the objective laws governing the evolution of science.

So much on the practical aspect of our endeavour. On the methodical aspect the following may be said in advance. The evolutionary laws for physics must be abstracted from the historical development of this science: there is no other source. The question arises whether the abstracted laws will still hold for the future development. In principle, new evolutionary laws may become operative and old ones may loose their power. Yet experience has shown that all laws gained by scientific abstraction have a far wider range of validity than that of its empirical origin. Hence we may expect that evolutionary laws, abstracted from a historical development extending over 300 years, will still hold for the next 50 or 100 years.

Before the method of scientific abstraction can be applied, the empirical material, viz., the physical theories, have to be subjected to logical analysis. Above all, they have to be freed of their historical 'egg-shells' [Lenin] which they have carried about for a rather long time. This preparatory work is absolutely necessary, but by no means easy. This may be the reason why historiography has been unable to discover evolutionary laws for physics.

Logical analysis plays again a decisive role in prognostics, i.e., in the [correct] application of evolutionary laws. There will hardly be time to comment on this.

1. THE GENERAL LAW FOR THE DEVELOPMENT OF PHYSICS

The development of physics, like any other genuine development, is dominated by the *law of contradiction*: removal of contradictions is the mainspring of evolution and the evolution itself is not straight but often contradictory. The latter holds in particular for the unification of the theoretical basis, as will be explained later (cf. 2.4). That the theoretical basis has to be corrected if it leads to contradictions with experience is generally recognized. Apart from such 'external' contradictions there may be 'internal' contradictions which are hints for the further development. Two kinds of 'internal' contradictions should be distinguished:

(a) logico-mathematical contradictions between *different* theories accepted at the same time, and

(b) contradictions turning up when a *single* theory is applied to particular problems.

Here are some classical examples:

(a1) Contradiction between Newton's *mechanics* and Maxwell's *electrodynamics*: the contradiction resides in the fact that the two kinematic invariance groups are incompatible since they correspond to $c = $ infinite and $c = $ finite ($c = $ limiting signal velocity [with respect to inertial systems]).

(a2) Contradiction between *thermodynamics* and Maxwell's *theory of the* [source-free] *electromagnetic field*: this contradiction manifests itself in the 'ultraviolet catastrophe' in the problem of spectral energy distribution of thermal radiation.

The contradiction (a1) has been removed by replacing Newton's by Einstein's mechanics while in case (a2) Maxwell's theory was replaced by the quantum theory of the electromagnetic field.

The contradictions of type (b) are not so well known. Examples are:

(b1) *cosmological paradoxes* in Newton's theory of gravitation;

(b2) *Gibbs' paradox* in classical statistical thermodynamics concerning the diffusion of two like gases; this paradox can also be considered as a logical contradiction between phenomenological thermodynamics and Boltzmann statistics;

(b3) the *exploding electron* in the Maxwell-Lorentz electrodynamics;

(b4) *Klein's paradox* in the original version of Dirac's theory of the electron: states of negative energy (mass) of a free particle;

(b5) the *divergencies* in quantum field theory.

Whether these are inconsistencies in the sense of logic or mathematics depends on the precise formulation of the theory concerned. The essential point is that in all these cases consistency can only be secured by either *limiting the range of application* in a way that appears arbitrary from the standpoint of the theory or else by *ad hoc assumptions* or *special methods* that have to be regarded as partial anticipation of a profounder theory.

(b1) has been overcome by Einstein's theory of gravitation, (b2) by quantum statistics. Resolution of (b4) led to the prediction of the positron (the first antiparticle) and to the discovery [W. Pauli] that in *any* relativistic quantum theory the number of particles does not satisfy a conservation theorem so that the particles should be considered as excited states of the field. Finally, (b3) re-emerges in modified form in quantum electrodynamics where it is removed (or rather: made innocuous) by renormalisation.

The interesting thing about these internal contradictions is that they do not restrict the usefulness of the theory in any essential measure while at the same time making it clear that the theory is fundamentally deficient.

Thus the internal contradictions of type (b) are neither logical inconsistencies *sensu strictu* nor dialectical contradictions reflecting fundamental properties of matter; rather, they express the imperfection of the theoretical knowledge attained, and thereby hint at a higher level of knowledge. In this resides their prognostic value.

If you like, you may count among the internal contradictions logical imperfections in the mathematical structure of the theory concerned. Two examples may serve as illustrations:

In Galilean-Newtonian kinematics you can apply a Galileo transformation to *any* frame of reference to obtain an equivalent frame. Thus, there exist an infinite number of Galilean-equivalence classes of frames none of which is kinematically distinguished. It is only by the general law of motion (dynamics) that one of these classes becomes a distinguished or preferential class, called the class of inertial frames. Thus, we have here a certain *discrepancy between kinematics and dynamics*. In the final analysis, it is this discrepancy which Huygens and Leibniz objected to. The discrepancy has been removed by the *c*-theory (Special Theory of Relativity), but not – as was desired by Huygens and Leibniz – by extending dynamical equivalence to all frames, but by restricting kinematic equivalence to the preferential class of Newton's inertial frames; only in these frames do particle kinematics and light kinematics coincide.

A second example is provided by the fact that the canonical commutation rules of h-c-theories (quantum field theories) admit *inequivalent representations*[1]; this makes it questionable whether the theory has a well-defined physical content.

2. Specific Laws for the Development of Physics

2.1. *The Trend Towards 'Abstractness'*

First, I mention the fact, particularly emphasized by MAX PLANCK[2], that the basic concepts of physics become more and more 'abstract', i.e., more and more removed from sense perception and 'everyday physics'. Planck remarks quite correctly that this development is a striking refutation of positivism and becomes understandable only as an approximation to the real world.

2.2. *Reduction of Properties to Processes; Modes of Reaction*

Another evolutionary law is the growing reduction of phenomenological properties to processes. Thus, the kinetic theory explains gas pressure as a result of momentum transfer from the molecules to the walls on reflection. Somewhat more complicated is the explanation of the colour of a non-luminescent body in terms of selective absorption and reflection of the incident light.

This last example may be used to introduce the important concept *'mode of reaction'*; by this we mean a property of a material system that manifests itself only in interaction processes. Thus, what we usually call 'the colour' of a non-luminescent body is in fact its mode of reaction to white light.

Modes of reaction [also called 'dispositional properties'] are in general not subjected to the classical logic of predicates but to a generalized logic with semi-Boolean algebra.[3] This is the reason why the so-called 'observables' of quantum mechanics, which really mean modes of reaction, are represented by non-commuting operators; indeed, the projection operators into which the latter may be resolved are but a special representation of a semi-Boolean algebra.[3]

It follows that the so-called 'complementary mode of description' (Bohr) introduced by quantum mechanics is not to be considered as a particularity of this theory that would eventually disappear together with that theory but rather as a manifestation of a general law in the evolution of physics.

On the basis of this law it may be predicted that a future theory will explain the inert mass of the fundamental particles as their mode of reaction to the physical vacuum. First attempts in this direction can already be found in present theories.

The last remark leads over to a further law that may be called:

2.3. *Reduction of Contingent Data*

By 'contingent' is meant 'not determined by the theory'. Typical examples of contingent [values of] physical quantities are the material constants of macroscopic (phenomenological) physics such as indices of refraction, elastic constants, viscosities, and the rest masses of fundamental particles in quantum mechanics. Apart from the latter, these quantities have already been reduced to atomic quantities and universal constants by atomic theory.

Present research aims at the reduction of rest masses and coupling constants to universal constants. But this is not the only aim of a theory of fundamental particles. We want to know why the fundamental particles, known or as yet unknown, with their seemingly contingent combinations of mass, charge, spin, hyperspin, etc., and no other, exist.

Here, the *contingency of existence* is attacked. This is a novel and philosophically most remarkable trend; though originating from an old evolutionary law, it is actually transcending it.

A similar trend can be observed in cosmology and cosmogony. Here it is the contingency of boundary or initial conditions we are no longer satisfied with.

Finally, there are attempts like that by E. A. MILNE[4] to extend the anticontingency trend to the very laws of nature, viz., to deduce these laws from the material content of the universe. Though Milne's attempt appears to be premature, the actual development of physics seems to tend in a direction where the old metaphysical opposition of law and matter is replaced by some kind of dialectical union of the two. The deducibility of the law of [gravitational] motion from Einstein's field equations may be regarded as a first step in this direction. The discovery of parity violation [is another example since it] reminds us that the material content of our world is likewise asymmetrical with respect to particles and antiparticles. In praxis, the reduction of contingent data goes hand in hand with the

2.4. *Unification of the Theoretical Basis*

Kepler's laws were shown by Newton to be consequences of his law of gravitational motion. The theory of heat was [effectively] reduced to mechanics and statistics by the kinetic theory. The Maxwell-Lorentz theory reduced optics to electrodynamics. Quantum mechanics made the theory of chemical bonds part of atomic theory. Quantum field theory explains the empirical connection between spin and 'statistics' [viz., permutational parity], left unexplained by quantum mechanics. Similarly, Einstein's theory of gravitation yields the proportionality between inert and gravitational mass unexplained in Newtonian mechanics.

Contrary to undialectical or utopian wishes and expectations, the process of unification is not one of uniform convergence; again and again it is interrupted by the opposite tendency. This is mainly due to the fact that new domains of nature become accessible thanks to the progress in experimental and observational techniques. The history of physics knows many premature (utopic) unifications resulting from excessive speculation and defective research. Thus, the phlogiston theory invented a thermal substance for reducing heat phenomena to Newtonian mechanics. For half a century the best physicists and mathematicians tried to explain the phenomena of light and electromagnetism in terms of a mechanical aether theory. Even Einstein who had given much thought to methodology and epistemology of physics was misled by the success of his gravitational theory to attempt a 'unified field theory', viz. a geometrical theory representing both gravitation and electromagnetism – a program to which he devoted the last 30 years of his life and which today appears even less rational than it did at the time of its conception.

Discounting thermodynamics and the theory of elementary particles (which is only in the making) we still have four irreducible theories in present physics:

quantum mechanics (h-theory)
relativity theory (c-theory)
quantum field theory (h-c-theory)
gravitational theory (κ-c-theory).

In parenthesis the theories are characterized by the universal constants appearing in them. Though the gravitational constant κ merely plays the role of a coupling constant in Einstein's equations – similar to the electric charge e in the Maxwell-Lorentz equations – it has been awarded the status of a universal constant. The justification lies in the fact that gravitation – in contrast to all other kinds of interaction – has a *universal* character: all kinds of matter contribute to the matter tensor T_{ik} which in Einstein's equations is the source of gravitational potentials $[g_{ik}]$.

Quantum mechanics has to be considered as a fundamental [irreducible] theory as it *cannot* be obtained from quantum field theory (by the process $c \to \infty$, say).

Similarly, c-theory cannot be obtained from the c-κ-theory: if you put $\kappa = 0$ in the Einstein equations, the resulting equations characterize a type of space that, though more special than the general Riemann space, is still far more general than the pseudo-Euclidean (Minkowski) space of c-theory; in fact, its curvature tensor does *not* vanish identically.

By way of contrast, Newtonian mechanics is a limiting case of [[both quantum mechanics ($h \to 0$) and]][5] relativistic mechanics [c-mechanics] for $c \to \infty$; furthermore, Newtonian gravitation can be obtained from the Einstein theory in the limit $c \to \infty$.[6] Thus, Newtonian mechanics is not an irreducible theory in the present meaning of this term.

If the above scheme is extrapolated the next step should be a

h-κ-c-theory

i.e., a generalized quantum field theory comprising gravitation. In view of the universal character of gravitation such a theory would probably lead to a quantization of the chrono-geometrical metric[7], with the Planck constant

$$l_\mathrm{P} = \sqrt{hc\,\kappa} \approx 10^{-33} \text{ cm}$$

in the role of a critical length.

The actual development of physics has taken a different course. On the one hand, the facts of elementary particle physics provide no evidence that gravitation plays a noticeable role in it, on the other hand they show clearly that traditional quantum field theory is insufficient. Any fundamental improvement of quantum field theory appears however connected with the introduction of a new fundamental constant of length

$$l \approx 10^{-13} \text{ cm.}$$

The next step would then be a

l-c-h-theory

as a theory of elementary particles without gravitation. Heisenberg's nonlinear spinor theory is an example of what such a theory may look like. To include gravitation, a further step would be necessary leading to a

κ-l-c-h-theory.

Such a theory would contain two universal constants of length. This may look strange but it is not a logical objection: the two constants would have entirely different physical meanings. Moreover, following Dirac and other scientists we have good reasons for assuming that the gravitational constant κ and hence l_p is itself a field [space-time] function so that its value may change, as compared with other length standards, in the course of cosmological evolution (expansion of the universe). Summarizing, we obtain Figure 1.

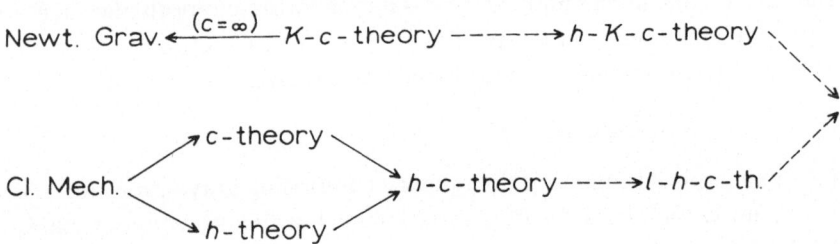

Figure 1 takes account of the universal constants only. Other schemes for the development of physics may be obtained when other points of view are taken as basic; but none of them appears equally fundamental.

NOTES

* Translated from 'Entwicklungsgesetze und Perspektiven der Physik', *Monatsberichte der Deutschen Akademie der Wissenschaften zu Berlin* **9** (1967) 538; paper read at the Symposion 'Dialectics and Modern Natural Science', held in Moscow from 26th to 29th October, 1966.
[1] Van Hove, *Physica* **18** (1952) 145; R. Haag, *Kgl. Danske Vid. Selsk. Mat.-Fys. Medd.* **29** (1955) No. 12; D. Hall and A. S. Wightman, *Kgl. Danske Vid. Selsk. Mat.-Fys. Medd.*

31 (1957) No. 5; A. S. Wightman and S. Schweber, *Phys. Rev.* **98** (1955) 817; H. Fukutome, *Prog. Theor. Phys.* **23** (1960) 989.

[2] M. Planck, *Das Weltbild der neuen Physik*, 12. Aufl., Leipzig 1953.

[3] M. Strauss, *Sitz.-Ber. Berl. Akad. Wiss., phys.-math. Kl.* (1936) 382; F. Kammer, *Nachr. d. Akad. Wiss. in Goettingen*, II. Math.-phys. Kl. Nr. 10 (1964).

[4] E. A. Milne, *Kinematic Relativity*, Oxford 1948.

[5] [Contrary to popular opinion, classical mechanics is *not* a limiting case of quantum mechanics. Cf. the paper on intertheory relations, this volume, Chapter XXII.]

[6] A. Einstein, *Berl. Ber.* (1916) 688; K. O. Friedrichs, *Math. Ann.* **98** (1928) 566; A. Trautman, *C. R. Acad. Sc.* **257** (1963) 617; G. Dautcourt, *Acta Physica Bolonica* **25** (1964) 637.

[7] H.-J. Treder, *Mber. Dt. Akad. Wiss. Berlin* **6** (1964) 88.

CHAPTER III

THE HUYGENS-LEIBNIZ-MACH CRITICISM IN THE LIGHT OF PRESENT KNOWLEDGE*

This paper has a double aim: first, to present a case study of the interdependence of the logic and the philosophy of science; and second, by so doing, to correct present day historiography.

Many historians and philosophers of science, and even some physicists, still consider Einstein's theories as a sort of implementation of the criticism levelled against Newtonian mechanics by HUYGENS[1], LEIBNIZ[2], and MACH[3]. As for Mach, this opinion goes back to EINSTEIN[4], as for Huygens and Leibniz, it goes back to REICHENBACH[5]. I wish to show that Reichenbach's verdict has to be substantially qualified while the Mach-to-Einstein story proves to be a myth, except for psychological matters.

To avoid misunderstanding, let me say that I am not concerned here with the general philosophy of either Leibniz or Mach, nor with the philosophical explications of Newton and *their* criticism. I am solely concerned here with physical theories, their criticism, and their logical relations – for details of proof I beg to refer you to a forthcoming paper in Synthese.[6]

I

I start with presenting the points in Newton's theory which appear open to objection to-day.

The first point is the *discrepancy between kinematics and dynamics*. I refer to the mathematical fact that the invariance group of Newton's dynamics is but a small subgroup of the full kinematic group of Newtonian space-time: the former depends on 10, the latter on infinitely many parameters. This results from the fact that Newtonian space-time is the direct product of space and time. In physical terms this means that in Newton's theory *all physical frames are kinematically equivalent irrespective of their relative motions*, while only those of a small preferential subclass, called *inertial frames*, are also *dynamically* equivalent.

The second point open to objection is the *unpredictability of the preferential class of inertial frames*. To be sure, this does not mean that the inertial frames are independent of the distribution and motion of matter in the universe as wrongly implied by Einstein in his famous thought experiment with the two rotating spheres; it just means that the *identification of the inertial frames is left to experience.*

The third and last point is the *discrepancy between Newton's general dynamics and his theory of gravitation.* The former implies the existence of *global* inertial frames, while the latter admits *local* inertial frames only, except in empty universe. This follows from the two fundamental properties of gravitation, viz., its universal character and its nichtab-schirmbarkeit.

I do not count the unexplained equality of inert and gravic mass among the objectionable points: otherwise every contingent feature of a theory would be objectionable. In fact, this equality is far less funda-mental than the two other properties which anyhow call for a geometrical theory of gravitation.

Now the first discrepancy – that between kinematics and dynamics – has already been removed by Einstein's Special Theory: the 10-parameter invariance group of Minkowski space-time, known as Lorentz or Poincaré group, is also the invariance group of relativistic dynamics. The other two objectionable points have only been removed by Einstein's General Theory.

Note that this has been achieved, not by granting dynamical equi-valence to all frames, but in precisely the opposite way, viz., by *re-stricting* kinematic equivalence, first to a *single* global uniform motion equivalence to be identified with Newton's inertial frames, and then to *local* inertial frames only. If you like you may see in this a partial return, on a higher level, to the space conception of Aristotle.

Now we are in a position to answer the questions at issue.

It seems that neither the second nor the third objectionable point had even been noticed by either Huygens or Leibniz. If this is correct none of them can be regarded a forerunner of Einstein's *General Theory.*

As for the first point – the discrepancy between kinematics and dynamics – both Huygens and Leibniz did object. Yet they did not anti-cipate the correct solution; in fact, they advocated dynamical equivalence of all frames, which implies Newtonian space-time. Thus it would be misleading to call them forerunners even of Einstein's *Special Theory,* though their criticism may be said to have paved the way for it.

Reichenbach's mistake in this matter is due to his acceptance of Einstein's identification of *general covariance* – a logical necessity like invariance under change of units – with *'general relativity'*, i.e., with the dynamical equivalence of all frames – a mistake in semantics first pointed out by KRETSCHMANN[7] and later by FOCK[8].

As for Mach, the story is quite different. Mach was well aware that he was not criticizing Newton's *physical theory* but merely his *way of presenting it* and, of course, his *philosophical explications*. In particular, he considered 'force' and 'inertial frames' as auxiliary concepts that should be eliminated from the fundamental equations. If this is done with the gravitational force, the resulting equations contain indeed only quantities that are invariant under the full kinematic group of Newtonian space-time. Thus, 'general relativity' (equivalence of all frames) is here achieved in a trivial way. At the same time, the resulting equations permit a physical interpretation of Newton's inertial or pseudo-forces in terms of true gravitational forces from distant masses. This is certainly an important advance in the understanding of Newton's theory, at least for the empiricist, but it does not entail any fundamental objection against the theory. Incidentally, the Mach transcription of Newton's theory turns out to possess *less* physical content than the original version, except for the interpretation of inertial forces; e.g., the whole of collision theory drops out. Thus, it cannot be taken as a substitute for Newton's original version.

Thus, Mach cannot be considered a forerunner of Einstein's *General Theory* either. It was only by an ingenious misinterpretation of Mach's writings that Einstein conceived of what he called 'Mach's Principle'. Even this principle has no part in the final theory, though it has certainly played a stimulating role in Einstein's search for a new theory of gravitation.

Sometimes a connection is seen between Mach's empiricism and Einstein's *Special Theory*. Einstein's definition of simultaneity is quoted as proof. Operationalists have made a whole philosophy out of this. As I have dealt with this aspect elsewhere[9] I confine myself to the following points. Einstein's analysis of simultaneity shows that *extended* or *frame time*, as distinct from *local time*, is a *kinematic concept* related to *velocity*; this is a true discovery in the logic of science. Yet it does *not* establish any *logical priority* nor the need for operational definitions. Moreover, Einstein's operational definition of simultaneity, though useful for practical purposes, is not wanted in a rational reconstruction of the theory since the latter is quite independent of the existence of light. Indeed, relativistic dynamics *admits* but does *not imply* the existence of zero mass particles [such as photons].

Thus, logico-mathematical analysis leads to the inescapable conclusion that Mach was perfectly right in declining to be considered a forerunner of Einstein's theories, either *Special* or *General*.

So much for the correction of historiography by logical analysis. Methodological analysis, which I have no time to present, merely confirms this conclusion.

II

I now turn to the inverse question, viz., the bearing of *philosophy* or *history of science* on the *logic of science*. Let me say in advance that I regard philosophy and history of science as inseparable: I take *philosophy of science* to mean the sum total of the lessons to be derived from the *actual history of science*.

Now the objection to the *first discrepancy* in Newtonian theory cannot be justified on purely logical grounds. *Logic* does not forbid a restriction of the invariance group if we pass from kinematics to dynamics. However, the objection is justifiable on *philosophical grounds*. Any materialistic conception of space and time requires that the properties of physical space-time be derived from, or related to, the properties of physical matter. The NOETHER theorem[10], which connects symmetries with conservation laws, permits us to give a precise formulation: *the properties of physical space-time should be identical with, or uniquely related to, the properties of properties of matter.*

Now the symmetries concerned in the applications of the Noether theorem are not primarily the symmetries of *space-time* but of those of *dynamics*. Hence *the materialistic conception of space-time demands that the space-time invariance group be a subgroup of,* (or identical with) *the dynamic invariance group,* and not the other way round as in Newton's theory.

In this sense the Huygens-Leibniz criticism is philosophically justified.

III

In conclusion I suggest that all disciplines of metascience be integrated into a self-correcting cybernetic system with information feedback between any two of the integrated disciplines. Such a self-correcting system we may call *theory of science*.

I hope my paper has made a modest contribution towards this end.

NOTES

* Paper read at the *Third International Congress for Logic, Methodology, and Philosophy of Science*, Amsterdam, 25 Aug.–2 Sept. 1967.
[1] 'Leibniz' Briefwechsel mit Huygens' in *G. W. Leibniz-Hauptschriften zur Grundlegung der Philosophie* (ed. by E. Cassirer), Leipzig 1904.
[2] *The Leibniz-Clark Correspondence* (ed. by H. C. Alexander), Manchester 1956.
[3] E. Mach, *Die Mechanik in ihrer Entwicklung, historisch-kritisch dargestellt*, Leipzig 1883.
[4] A. Einstein, 'Die Grundlage der allgemeinen Relativitätstheorie', *Ann. d. Phys.* 49 (1916) 771–814, §2; 'Ernst Mach', *Phys. Z.* 17 (1916) 101–104; 'Prinzipielles zur allgemeinen Relativitätstheorie', *Ann. d. Phys.* 55 (1918) 241.
[5] H. Reichenbach, 'Die Bewegungslehre bei Newton, Leibniz und Huyghens', *Kantstudien* 29 (1924) 416.
[6] M. Strauss, 'Einstein's Theories and the Critics of Newton', *Synthese* 18 (1968) 251–284. [Chapter 15 of this volume.]
[7] E. Kretschmann, 'Über den physikalischen Sinn der Relativitätspostulate', *Ann. d. Phys.* 53 (1917) 575.
[8] V. A. Fock, 'Three Lectures on Relativity Theory', *Rev. Mod. Phys.* 29 (1957) 325. *Theorie von Raum, Zeit und Gravitation*, Berlin 1960.
[9] E. Noether, 'Invariante Variationsprobleme', *Göttinger Nachr.* 1918, 235–257.
[10] M. Strauss, 'Ist die Isotropie der Lichtausbreitung in einem Inertialsystem eine Konvention?', *Mber. Dtsch. Akad. Wiss. Berlin* 7 (1965) 364–365.
'The Lorentz Group: Axiomatics – Generalizations – Alternatives', *Wiss. Z. Friedrich-Schiller-Univ. Jena*, Math.-Nat. R. 15 (1966) 109–118. [Chapter 14 of this volume.]

CHAPTER IV

MAX PLANCK AND THE RISE
OF QUANTUM THEORY*

If one compares the biographical, historiographical, and philosophical literature about Max Planck with that about Albert Einstein the relative paucity of the former is quite conspicuous. One of the main reasons for this state of affairs is no doubt the historical coincidence that even more brilliant stars soon appeared on the physical scene. Another reason may be found in the personality of Planck himself. Yet there seems to be still another reason which is of a more general nature, being related to sociology rather than to physics. Planck protrudes into our century as a representative of that older generation of physicists for whom materialism in natural science was a matter of course. Hence the various idealistic trends appearing at the turn of the century and affecting the general intellectual climate in increasing measure could not claim Max Planck for their cause. If in recent years the interest in Planck's scientific work has become more pronounced, one of the main reasons is no doubt the fact that more and more scientists are turning away from *positivism* which had always been opposed by Planck. In a strictly physical sense, we are today likewise in a much better position to judge the work of Max Planck; most of his problems have meanwhile been solved by the *quantum theory of fields* in a more satisfactory manner than has been possible by the old ['classical'] quantum theory or by nonrelativistic quantum mechanics. In view of this the question may be raised whether Planck's aversion to quantum mechanics, shown up to the end of his life, is solely due to his maintaining the metaphysical ideal of deterministic causality or perhaps to other, and maybe progressive, reasons.

This question will not be considered in the following; its investigation, though desirable, does not promise any new results. Also the important question as to Planck's position in the controversial history of statistical thermodynamics and its foundation will only be touched on occasionally, particularly since this question has already been discussed in a way likely to provoke further investigations[0].

Our main theme is the *genesis* of quantum theory, and our main interest in this historical process is focused on the same aspects that dominated the thinking of Planck: the *striving for a unified world picture* and the *mutual relations between the physical theories* constituting this picture.

Indeed, it was this striving that led Planck, contrary to his own intention and expectation, to results that initiated a revolution in the conceptual foundations of physics. This presents a further aspect of general validity: the *existence of objective laws for the evolution of physics* that assert themselves in spite of subjective errors, and often with their very help.

I

When, in 1889, Max Planck was appointed to succeed Kirchhoff at Berlin University hardly anyone could foresee what glorious fame he would bestow on the *alma mater berolinensis*. Indeed, his Munich dissertation [doctoral thesis] of 1879 passed almost unnoticed although it contained the first formulation of the Second Law of Thermodynamics that could claim general validity.

As I know for sure from conversation with my university teachers, none of them had an understanding of its content. They let the dissertation pass as doctor thesis solely, so it seems, because they knew me from my other work in Practical Physics and in the Mathematical Seminary. Even those physicists who were more familiar with the topic showed no interest, let alone approval. Helmholtz doesn't seem to have read the paper at all, and Kirchhoff disapproved of the content *expressis verbis* with the remark that the concept of entropy cannot be applied to irreversible processes since its magnitude could only be measured, and hence defined, by a reversible process [1]

– a remark that must have given Planck a bitter foretaste of the opposition to his work from the side of positivism; it also brought Planck, right at the beginning of his career, into *personal* opposition to the representatives of this mistaken doctrine.

Not much better was the fate of the following work which was devoted to a deduction of the equilibrium conditions for multiphase systems or gas mixtures from the Second Law and which proved the author to be a master of thermodynamics.

Everywhere fruitful results turned up. Unfortunately, as I discovered afterwards, the great American theorist Josiah Willard Gibbs had obtained the same results before me ... so that in this field, too, external success was denied to me.[2]

At what time Planck became acquainted with the work of GIBBS[3] is not precisely known. Anyhow, a reference to Gibbs' work can be found not only in his 1883 paper on the thermodynamic equilibrium of gas mixtures[4], but already in a previous paper[5] which was finished in December, 1881. This may shed some light on a question we shall discuss later, viz., the question whether Planck had any knowledge of Rayleigh's 1900 paper when establishing his radiation law by the famous 'interpolation', or whether he used then only the experimental results obtained by Kurlbaum and Rubens.

The following papers, which appeared under the common heading "Ueber das Prinzip der Vermehrung der Entropie", did not bring any new results in principal questions that would go beyond the classical work of Gibbs, though Planck's less abstract mode of presentation may have contributed to a more general understanding of the theory.

In view of all this, Planck's appointment to a chair at Berlin University may be ascribed above all to Helmholtz' keen sense for creative talents, which had proved true before in his promotion of Heinrich Hertz.

II

To a thinker less profound and less thorough the world picture devised by theoretical physics before the turn of the century might have appeared as being quite satisfactory. Ponderable matter was taken care of by *Mechanics*, the electromagnetic field by *Maxwell's Theory*, and everything connected with heat by *Thermodynamics*. But what about the coherence of these parts? True, there was the *Kinetic Theory of Heat*, unifying Thermodynamics with Mechanics, if one accepted the statistical interpretation of entropy, which Planck was not yet prepared to do. What about the connection between *Thermodynamics* and *Maxwell's Theory*? Some connection was bound to exist, in view of the existence of thermal radiation. Yet the theory of thermal radiation had not advanced beyond Kirchhoff's law and WIEN's displacement law[6]. True, Wien had also suggested a formula for the spectral distribution of black-body radiation which agreed quite well with the measurements in the temperature and frequency region investigated; yet its theoretical foundation was far from being secure. This gap in the theoretical world picture Planck resolved to close; the closing arch, connecting Thermodynamics with Electrodynamics, should become the crowning element completing the whole structure of theoretical physics. (That still another element, connecting

Mechanics and Electrodynamics, was missing Planck only realized after Einstein's 1905 paper, but then he was one of the first to do so.)

Today, Planck's program, as well as its first implementation, appears to us as a strange mixture of insight and error, acuteness and naivity, rarely found in the history of physics.

Let us start with the program. Planck correctly realized – and that is his proper merit in this question – that a theoretical analysis of thermal radiation was bound to reveal the connection between Thermodynamics and Electrodynamics. He did not see at that time that the problem of the spectral distribution of radiant energy in thermal radiation was a problem not of phenomenological but of *statistical* thermodynamics, as should have been obvious from the analogous problem of velocity distribution in a gas – in either case the distribution variable is a *microphysical* quantity not occurring in [phenomenological] thermodynamics. That Planck did not realize this was his first and fundamental mistake.

Planck did not see either that Maxwell's theory was bound to lead to Rayleigh's radiation formula and hence to the 'ultraviolet catastrophe' *for purely dimensional* [7] *reasons* [8], and hence quite independently of any statistical interpretation of thermodynamics. In other words: *Maxwell's Theory is incompatible with Thermodynamics*, and hence there cannot be any link connecting the two in a coherent way. Thus Planck's striving for a coherent and unified world picture was bound to lead, with logical and historical necessity, to precisely the reverse of the desired aim, namely to the discovery of the *contradiction* between the two theories and subsequently to the removal of this contradiction by quantum electro-dynamics. It is a hypothetical question to ask what Planck would have done had he become aware of this contradiction at the time he conceived his program. The early discovery of the contradiction would probably have split the community of physicists into two camps – just as the discovery of certain inconsistencies in quantum field theory has done in our day – one with the maxim 'Back to the mechanical aether theory of light', the other with the maxim 'Forward beyond Maxwell'. It is idle to speculate which camp Planck would have joined.

Last but not least, Planck, impressed by the first experimental results obtained by Paschen and by Lummer and Pringsheim, was convinced at that time that Wien's radiation formula was theoretically correct (in the sense of Maxwell's theory of course). Yet the analysis of Wien's argument shows quite clearly that the whole argument rests on the assumption that the radiation frequency depends only on the velocity of the emitting atom. Consequently, in the case of thermal equilibrium [between atoms and radiation] the Maxwell distribution is simply transferred from the atomic velocities to the radiation frequencies.

Indeed, Wien's radiation formula becomes identical with the Maxwell distribution if transcribed from wavelengths to frequencies (s. formula (7)) and if the identification

$$akv = bv = mv^2/2 \tag{1}$$

is used. Thus, Wien's assumption amounts to the statement that the thermal radiation behaves like a corpuscular [ideal] gas of particles having the energy bv – a statement absolutely incompatible with Maxwell's theory. This fact, too, escaped Planck's attention; it was discovered much later by Einstein in the context of his analysis of fluctuation phenomena.

III

We now turn to the implementation of Planck's program.

The first step was of course to bring the program into a form accessible to mathematical treatment.

The obvious idea would have been to use some atomic model, to calculate the interactions between these free atoms and the Maxwell field, and to establish the condition of equilibrium ('emission = absorption'); this method, already used by Wien, was later employed by Einstein for deducing Planck's formula. Planck however was not prepared at that time to use a method involving the statistical theory of heat applied to atoms. Instead, he considered the interaction of a *single* atom, fixed in space, with the radiation field, hoping the interaction would turn out to be irreversible; whatever its initial state, the system would then pass eventually into a stationary state in which the radiation would have the energy distribution looked for. Since, according to the Kirchhoff theorem, the choice of the atomic model was in no way restricted, Planck choose a Hertz oscillator – a choice that was to turn out to be very fortunate in two respects.

In the first place, the mean energy $\bar{U}$ of an oscillator of frequency v turned out to be related to the spectral energy density u by the equation

$$u = (8\pi v^2/c^3)\ \bar{U}. \tag{2}$$

Now the factor

$$N = 8\pi v^2/c^3 \tag{3}$$

is precisely, as first shown by Rayleigh (1900), the number of different modes of vibration per unit volume in a cavity with reflecting walls. In other words: *Planck's oscillator may be regarded as an electromagnetic eigenvibration of the cavity.* In line with this, Planck himself emphasized in his lectures given in the twenties that it does not matter whether his oscillator be thought of as an oscillating electric charge or else as an electromagnetic eigenvibration of the cavity.[10]

The second reason why Planck's choice was fortunate will be explaines later (p. 44).

Another result of rather general importance, re-established by Planck in this first phase of his investigations, is the oscillator equation

$$\frac{dU}{dt} = Z \frac{df}{dt} - (\pi/3)\ N \left(\frac{df}{dt}\right)^2 \tag{4}$$

where U is the oscillator energy [as a function of time t], f the dipole moment of the oscillator, and Z the component of the electric field in the dipole direction. Now the second [subtracted] term on the right-hand side is independent of Z and positive definite; hence it represents the [so-called] *spontaneous emission*. The first term on the right-hand side can be positive or negative, as correctly noticed by Planck, depending on whether the directions of Z and df are parallel or antiparallel. Hence the first term represents both the *induced emission* and the *induced absorption*. We have here a complete parallel to Einstein's 1917 deduction of Planck's law where likewise all three processes mentioned are taken into account and where the coefficients (probabilities) for induced emission and induced absorption are taken to be equal.

IV

The most interesting conclusion arrived at by Planck in the course of these investigations however concerns the *statistical character* – or, to use Planck's expression, the *natural disorder* – of thermal radiation itself. The [somewhat involved and strange] story of this is as follows.

As already mentioned above, Planck had hoped that in a consistent application of Maxwell's theory the interaction between a [material] oscillator and the radiation field would turn out automatically to be irreversible. Indeed, in his First Communication [to the Berlin Academy of Sciences] (1897) Planck regards it as "the fundamental task" of theoretical physics "to reduce irreversible changes (einseitig verlaufende Veränderungen) to conservative actions (conservative Wirkungen)" – a task, he adds, that has not at all been solved in a satisfactory manner by the kinetic theory, as Zermelo (1896) had just shown. In the case of the electromagnetic field, however, the situation, as seen by Planck, would be more favourable: "Einen lediglich aus conservativen Wirkungen bestehenden und dennoch einseitig verlaufenden Vorgang glaube ich erblicken zu müssen in den Wirkungen, welche ein ohne inneren Reibungs- und Leitungswiderstand schwingender Resonator auf die ihn erregende Welle ausübt." Much though this belief testifies to Planck's ability to think in the categories of dialectics, he soon had to convince himself of its untenability in view of an objection made by Boltzmann and corresponding exactly to Zermelo's *umkehreinwand*. Indeed, just as the equations of mechanics are invariant under time reversal, Maxwell's equations are invariant under simultaneous inversion of time and magnetic field vector so that the processes of emission and absorption could also be inverted.

In this situation Planck felt compelled to fall back on the method used by Boltzmann in establishing the H-theorem – to use the more radical method of statistical mechanics he was still not prepared. Thus the immediate aim was now to construct an expression involving only electromagnetic quantities that would be an increasing function of time and hence be acceptable as representing the entropy of the radiation field. To exclude any objection based on time inversion it was further necessary to formulate a hypothesis corresponding to Boltzmann's hypothesis of molecular disorder and concerning the irregular nature of thermal radiation with such care that its compatibility with Maxwell's equations could not be doubted.

In trying to solve this problem Planck choose a way that was quite original but wrong in principle; nevertheless it proved heuristically useful. Instead of transferring [as would have been the obvious thing to do] the disorder hypothesis from the ponderable to the imponderable

energy carriers, i.e., from the molecules to the electromagnetic eigenvibrations of the cavity, Planck starts from the experimental fact
characteristic of atomic radiation, that even the sharpest spectral line
has a finite breadth corresponding to a finite frequency interval. This
offered the possibility, used by Planck, of locating the disorder in the
amplitudes and phases of the Fourier components (partial waves) of
such atomic radiation. In carrying out this idea a strict distinction had to
be made between slowly and rapidly changing quantities as only the
latter had to be averaged – a method anticipating Born's masterly treatment of similar problems. This averaging procedure, applied to (4),
finally led to the 'fundamental equation'

$$\frac{d\bar{U}}{dt} = -2\sigma v\bar{U} + \frac{3\sigma v}{2\pi N} I \tag{5}$$

where σ is the damping factor (logarithmic decrement) of the oscillator
and I the (slowly changing) mean value of the radiation intensity of frequency v at the place of the oscillator. The second term on the right-
hand side represents induced absorption while the first term represents
spontaneous emission. The absence of a term representing induced
emission shows clearly that equation (5) does not refer to elementary
processes but to certain averages.

In the stationary case I is independent of t and the integration of (5)
yields the simple relation

$$I = (4\pi/3)\, N\bar{U} \tag{6}$$

which is equivalent to the previous equation

$$u = N\bar{U} \quad (N = 8\pi v^2/c^3). \tag{2}$$

V

In order to understand the further course of Planck's considerations and
their proper significance it is necessary here to point out a characteristic
difference between Planck's interpretation of Equation (6) and the interpretation current today. In the current interpretation, which goes back
to Rayleigh, $\bar{U}$ is the *canonical* or *microcanonical* average, in the sense of
statistical thermodynamics, of the *radiation* energy belonging to a single
mode of vibration of frequency v; Equation (6) or (2) is then completely

trivial since it follows from the definition of the quantities concerned. In contrast to this, Planck's $\bar{U}$ does not refer to radiation but to a *ponderable fixed oscillator* merely *exchanging* energy with the radiation field. Furthermore – and this is the decisive difference – the Planck average is obtained by averaging over the 'rapidly' changing parameters of the oscillator that correspond to amplitudes and phases of the partial waves (Fourier components) of "natural" radiation, with the Planck average serving as a [theoretical] definition of 'natural'.

Planck was fully aware of this fundamental difference:

Das Prinzip der Unordnung, auf welchem jede Art der Irreversibilitaet zu beruhen scheint, liegt bei der Gastheorie in einem ganz anderen Moment als bei der Waermestrahlung. In den Gasen sind es die zahlreichen ponderablen Molecuele, welche durch die Unregelmaessigkeit ihrer Lage und Geschwindigkeit die Unordnung bedingen; im durchstrahlten Vacuum dagegen sind es die zahlreichen Strahlenbuendel, welche durch ihre unregelmaessig wechselnde Schwingungszahl und Intensitaet zur Bildung der Entropie Veranlassung geben. Bei den Schwingungen eines einzelnen Resonators kommt diese Unregelmaessigkeit ebenso gut zum Ausdruck wie bei der Strahlung im freien Raum. ... Dementsprechend repraesentiert die stationaere Schwingung eines in einem stationaeren Strahlungsfeld befindlichen Resonators mit bestimmter Eigenperiode nicht etwa einen einheitlichen Elementarvorgang, d.h. eine einfache Sinusschwingung mit constanter Amplitude und Phase ..., sondern sie besteht in einer Ueberlagerung sehr vieler kleiner Einzelschwingungen mit nahezu gleichen Perioden und constanten Amplituden und Phasen, oder auch, was mathematisch genau auf das Naemliche hinauskommt, in einer einzigen Schwingung mit constanter endlicher Amplitude, aber unregelmaessig veranderlicher Phase. In jedem Fall kann man von einer Unordnung, also auch von einer Entropie und einer Temperatur des Resonators sprechen.[11]

[Having thus ruled out any possible objection based on time inversion] Planck proceeds to show that the entropy expressions corresponding to Wien's radiation law

$$u(v, T) = N(v)\, \bar{U}(v, T) = Nbv \exp(-av/T) \qquad (7)$$

i.e., the expressions

$$S = -(\bar{U}/av) \log(\bar{U}/ebv) \qquad \text{(for the oscillator)} \qquad (8a)$$

and

$$s = -(u/av) \log(u/eNbv) \qquad \text{(for the radiation density)} \qquad (8b)$$

satisfy, just like Boltzmann's expression for $-H$ in kinetic gas theory, the principle of growing entropy. This completes the proof that Wien's radiation formula is compatible with thermodynamics.[11a]

This result is not surprising [and indeed did not require any proof] in view of the fact that Wien's radiation formula satisfies Wien's displacement law and hence also the Second Law of Thermodynamics (as applied to reversible processes). Thus, the decisive question was not that of *admissibility* (compatibility with thermodynamics) but that of the *theoretical necessity* of Wien's radiation formula.

As to this critical issue, Planck remarks that any generalization of (8a) of the form

$$S = - \left(\bar{U}/f\,(v)\right) \log \left(\bar{U}/g\,(v)\right)$$

with arbitrary functions f and g would lead back to the old expression (8a) in view of Wien's displacement law, and that he has not succeeded in constructing any other expression satisfying the principle of growing entropy. From this he felt entitled to draw the mistaken conclusion that Wien's radiation formula is a necessary consequence of thermodynamics and Maxwell's theory.

VI

At this stage experiment intervened. Before Planck could be made to realize his mistake by Rayleigh's 1900 paper, new measurements by Lummer and Pringsheim in the region of larger wavelengths became known which showed considerable deviations from Wien's formula. This induced Planck to re-examine the problem. This re-examination was not only more profound but it also yielded an entirely novel and most remarkable result which was to become the starting point for Planck's famous 'interpolation' [and thus the last stepping-stone on the boggy road to the new radiation law].

The question which Planck now resolved to investigate was this: given the two Laws of Thermodynamics, what can be said about the *relation* between *energy* and *entropy* of *any* radiation? In order to answer this question Planck considers a *non-equilibrium* state in which the energy of the oscillator differs from the stationary value U by a [relatively] small amount ΔU. For the rate of change of the total entropy Σ Planck obtains the equation

$$\frac{d\Sigma}{dt} = - \frac{dU}{dt} \Delta U f\,(U) \tag{9}$$

with

$$f(U) = -\frac{3}{5}\frac{\mathrm{d}^2 S}{\mathrm{d}U^2}. \qquad (10)$$

The only thing that can be said about the [unknown] function $f(U)$ is that its value must be positive [for all positive values of U] so that $\mathrm{d}\Sigma/\mathrm{d}t > 0$ [increase of total entropy with time] because ΔU and $\mathrm{d}U/\mathrm{d}t$ are proved to be of opposite sign.

The significance of this result is as follows.

First – and this is main point concerning methodology – the extension of purely thermodynamic analysis to *non-equilibrium* states brings *no principal advance* in the spectral distribution problem beyond Wien's displacement law: *in either case a function of one variable remains undetermined*. This confirms the remark made above [Part II] that the spectral distribution problem is a problem in the statistical theory of heat and not a problem of pure thermodynamics. This rather negative result may appear paradoxical in view of the fact that a certain kind of statistical disorder *has* been taken into account by Planck. We shall resolve this apparent paradox later [cf. Part VIII].

Second – and this is the novelty – there appears for the first time in a fundamental investigation the *second derivative of the entropy with respect to the energy* which later served Planck as starting point for his famous 'interpolation'. The physical meaning [or rather significance] of this quantity is obvious from the above equations: the quantity determines the *speed with which a non-equilibrium state approaches the stationary state*. More generally, the named quantity is a *reciprocal measure of the thermal fluctuation of the local energy density*, as later shown by Einstein [1909].

Planck's paper quoted above contains another interesting result which is of fundamental importance for understanding [the theoretical implications of] Wien's radiation formula, anticipating as it does an equivalent result obtained much later by Einstein [1909].

In the hope of determining the unknown function $f(U)$, Planck considers the case that instead of *one* oscillator n completely similar oscillators are present all of which have the same initial energy $U + \Delta U$. The total change of entropy during the time $\mathrm{d}t$ is then obtained by replacing U by nU in formula (9):

$$d\Sigma_n = - n^2 \, dU \, \Delta U f \, (nU). \tag{11}$$

If the oscillators are assumed to be *completely independent* of one another the total change of entropy must be n times as large as if only *one* oscillator were present, i.e.,

$$d\Sigma_n = n \, d\Sigma = - n \, dU \, \Delta U f \, (U). \tag{12}$$

Comparison of the last two equations yields the functional equation for the unknown function f:

$$f \, (nU) = n^{-1} f \, (U) \tag{13}$$

which has the general solution

$$f \, (U) = \text{const.}/U \tag{14}$$

so that

$$\frac{d^2 S}{dU^2} = - \text{const.} \, U^{-1} \tag{15}$$

or [after integration]

$$S = - \, (U/A) \log \, (U/B), \tag{16}$$

where the integration constants A and B are functions of the frequency v only. If Wien's displacement law is taken into account, (16) gives precisely Wien's radiation formula.

Thus, *Wien's radiation formula is equivalent to the assumption that the radiation field does not interfere with the independence of Planck's oscillators*, or – if we identify these oscillators with the proper vibrations of the cavity-enclosed radiation field as suggested by Rayleigh's [1900] paper – *that the [local] fluctuations of the radiation energy density are completely incoherent* [i.e., independent of one another] as in the case of an ideal gas. It can only be ascribed to Planck's firm belief in the theoretical necessity of Wien's radiation formula that this obvious contradiction to the Maxwell Theory was not noticed by him.

Incidentally, Planck's assumption of an independent behaviour of his n oscillators *within the radiation field* is not only unwarranted theoretically but also in flat contradiction to the experimental facts of optical coherence without which there would be no interference phenomena. (That there exists a limit to the coherence of 'natural' [atomic] radiation, measured by the *coherence length*, belongs into a different context and was not even known at that time.)

To summarize [this pre-natal development of Planck's quantum theory]: Planck's original hope that the principle of growing entropy, applied to Maxwell's radiation field, would solve Kirchhoff's problem of spectral energy distribution has proved deceptive although the additional hypothesis of 'natural disorder' seemed to comprehend the statistical features of thermal radiation. Nevertheless, Planck's investigations proved useful and fertile. Not only did the results obtained facilitate a deeper insight into the problem but they may also have shaken Planck's belief in the theoretical necessity of Wien's radiation formula so that new theoretical or new experimental arguments could attract his attention.

VII

For Planck the new arguments came from the side of experimental physics. New experiments by Kurlbaum and Rubens in the region of still larger wavelengths not only confirmed the deviations from Wien's formula found by Lummer and Pringsheim but also showed clearly that the deviation becomes the larger the smaller the frequency and the higher the temperature. This induced Planck to use the freedom in the choice of the function

$$\frac{\mathrm{d}^2 S}{\mathrm{d} U^2} = -f(U)$$

and to choose a function f that would lead to agreement with the experimental results. In his original communication on this subject of October 19, 1900, he says:

Im Verfolg dieses Gedankens bin ich schliesslich dahin gekommen, ganz willkuerlich Ausdruecke fuer die Entropie zu construiren, welche, obwohl complizirter als der Wiensche Ausdruck, doch alle ‹ Anforderungen der thermodynamischen und elektromagnetischen Theorie ebenso vollkommen Genuege leisten wie dieser. Unter den so aufgestellten Ausdruecken ist mir nun einer besonders aufgefallen, der dem Wien'schen an Einfachheit am naechsten kommt, und der, da letzterer nicht hinreicht, um alle Beobachtungen darzustellen, wohl verdienen wuerde, naeher geprueft zu werden. Derselbe ergibt sich, wenn man setzt

$$\frac{\mathrm{d}^2 S}{\mathrm{d} U^2} = \frac{\alpha}{U(\beta + U)}. \tag{17}$$

Er ist bei weitem der einfachste unter allen Ausdruecken, welche S als logarithmische Funktion von U liefern (was anzunehmen die Wahrscheinlichkeitsrechnung nahe legt) und welche ausserdem fuer kleine Werte von U in den obigen Wien'schen Ausdruck uebergehen.

According to this report, Planck has used a typical method of *generalization*: to a linear term a quadratic term is added. No mention is made of any sort of interpolation. In particular, there is not the slightest hint at the circumstance that the added term corresponds to the radiation formula published a few months before by Rayleigh. The same is true of Planck's publications in the following years. His first paper[12] mentioning Rayleigh's formula appeared in 1910, but the paper referred to is not any of Rayleigh's papers but a paper by Jeans from the previous year. It is not until the Solvay Congress (1911) that Planck makes a direct reference to Rayleigh's fundamental paper of 1900.

In his later (1943) report 'Zur Geschichte der Auffindung des physikalischen Wirkungsquantums' there is likewise no evidence supporting the assumption Planck had known Rayleigh's formula when doing his 'interpolation' in 1900. This later report differs however from the original report as may be seen from the following quotation:

In der Sitzung der Deutschen Physikalischen Gesellschaft vom 19. Oktober 1900 teilte F. Kurlbaum die Resultate der von ihm in Gemeinschaft mit H. Rubens fuer sehr grosse Wellenlaengen ausgefuehrten Energiemessungen mit, aus denen unter anderem hervorging, dass mit steigender Temperatur die Strahlungsintensitaet des schwarzen Koerpers immer angenaeherter proportional der Temperatur T wird ... Da mir dieses Ergebnis schon einige Tage vor der Sitzung durch muendliche Mitteilung der Autoren bekannt war, so hatte ich Zeit, noch vor der Sitzung die Folgerungen daraus auf meine Weise zu ziehen ... Wenn fuer hohe Temperaturen T die Strahlungsintensitaet proportional der Temperatur wird, so ist ... auch die Energie des Oszillators ihr proportional, also:

$$U = CT,$$

und daraus nach (3) (i.e. $\mathrm{d}S/\mathrm{d}U = 1/T$) durch Integration:

$$S = C \log U.$$

Folglich:

$$\frac{\mathrm{d}^2 S}{\mathrm{d}U^2} = -\frac{C}{U^2}.$$

Diese Beziehung tritt also fuer grosse Werte von U an die Stelle der fuer kleine Werte von U gueltigen Beziehung (5), viz.,

$$\frac{\mathrm{d}^2 S}{\mathrm{d}U^2} = -\frac{A}{U}.$$

If one now looks for a general equation – so Planck continues – which would contain the last two equations as limiting cases the simplest equation would be Equation (17).

Thus, according to this account it looks as if Planck had *consciously* (and not only *de facto*) interpolated between two extremal cases.

A third version can be found in Planck's reply to the speeches made on the occasion of his eightieth birthday.[13] There he said:

Besonders um die Jahrhundertwende gab es kaum eine Sitzung, in der ich nicht zugegen war, und kaum eine Nachsitzung, die ich versäumt habe. Damals spielte sich unter anderem auch ein reger Meinungsaustausch ab über die Gesetze der Wärmestrahlung. An den Debatten darüber beteiligten sich hauptsächlich Lummer, Pringsheim, Jahnke, Thiesen, Kurlbaum, Rubens. Die verschiedensten Formeln wurden vorgeschlagen und diskutiert. Die schließliche Entscheidung brachten aber ... nicht die Versuche von Lummer ..., sondern die Messungen von Rubens und Kurlbaum, welche insbesondere zeigten, daß die Intensität eines monochromatischen Strahles bei hohen Temperaturen proportional der Temperatur wird. Ich erinnere mich sehr wohl, daß Rubens damals ... zu mir sagte: Wie es auch sei, das eine steht fest, die Intensität der monochromatischen Strahlung hat als einen Faktor die Temperatur und als anderen Faktor einen Ausdruck, der bei unbegrenzt steigender Temperatur endlich bleibt.

Unfortunately, this reminiscence does not allow either to draw any safe conclusion as to the course of events since their temporal order of sequence is left in the dark. Still, the context in which Planck uses the word 'Entscheidung' leaves hardly any doubt that he meant an experimental decision about [competitive] formulae already [tentatively] found by him, and not his own decision in favour of Equation (17).

Thus, there exists no convincing argument at all for the view that Planck's procedure was in any way different from what he has described it to be quite frankly in his original communication of October 19th, 1900. Of course, it is quite likely that Planck became aware of the interpolating character of his tentative Equation (17) very soon afterwards. The 1943 version of the story may easily be explained by Planck's intention to give, for good didactical reasons, a somewhat rationalized presentation expressing more clearly the physical significance of (17), rather than a mere repetition of his original more factual account. In any case it is quite clear that Planck did not know then the Rayleigh law.

It remains to be shown that Equation (17) does imply Planck's law if Wien's displacement law is taken into account. Integrating (17) one obtains

$$\frac{1}{T} = \frac{dS}{dU} = -\frac{\alpha}{\beta} \ln\left(\frac{\beta}{U} + 1\right), \tag{18}$$

and hence

$$U = \frac{\beta}{e^{-(\beta/\alpha T)} - 1} \qquad (19)$$

or

$$u = NU = \frac{8\pi v^2}{c^3} \frac{\beta}{e^{-(\beta/\alpha T)} - 1}. \qquad (20)$$

Since Wien's displacement law requires

$$u(v, T) = v^3 f(v/T) \qquad (21)$$

one has to put

$$\beta = hv \quad (h \text{ constant}). \qquad (22)$$

(Instead of the negative constant α one writes today $-\alpha = k$.)

The constant h is identical with the constant b in Wien's radiation formula

$$u_{\text{Wien}} = (8\pi v^2/c^3)\, bve^{-av/T}. \qquad (23)$$

Using this formula and the measurements by Kurlbaum and Paschen, Planck had calculated its value in a previous paper (March, 1900) when he was still convinced of the theoretical necessity of Wien's formula, obtaining

$$h = b = 6.885 \; 10^{-27} \text{ erg sec}. \qquad (24)$$

This may justify the name 'Planck's constant', though this name is apt to support the mistaken view that the constant had been introduced by Planck and its value first calculated on the basis of Planck's law [indeed, it would be more correct to call it the *Wien-Planck constant*].

Today we know that Planck's [factual, not conscious] interpolation has a very simple but deep-going physical significance. As shown by Einstein in his fundamental papers on the theory of thermal fluctuations, the reciprocal of the expression $f(U) = -d^2U/dS^2$ considered by Planck is a direct measure of the mean square deviation of energy in a partial volume at constant temperature and volume:

$$\overline{(\Delta U)^2} = k \frac{1}{f(U)} = -k \left(\frac{d^2 S}{dU^2}\right)^{-1}. \qquad (25)$$

Hence, Planck's interpolation formula (17) is equivalent to

$$\overline{(\varDelta U)^2_{\text{Planck}}} = \overline{(\varDelta U)^2_{\text{Wien}}} + \overline{(\varDelta U)^2_{\text{Rayleigh}}} \tag{26}$$

or, since the Wien and the Rayleigh law correspond to the classical corpuscular theory and the classical wave theory of light, respectively,

$$\overline{(\varDelta U)^2_{\text{Planck}}} = \overline{(\varDelta U)^2_{\text{corp}}} + \overline{(\varDelta U)^2_{\text{wave}}} \,. \tag{27}$$

Thus, *Planck's radiation law is a synthesis of the two classical theories of light which were in contest with one another for centuries.* However, as these classical theories merely represent the limiting cases $h\nu/kT \to \infty$ and $h\nu/kT \to 0$, respectively, it is more correct to call it a *dialectic uplation* (dialektische Aufhebung) of the two classical theories.

Thus, without knowing it Planck became the founder of the modern view concerning the 'dual' nature of matter which later found its adequate mathematical formulation in quantum field theory.

VIII

For the time being [1900–1909] the deeper theoretical significance of Planck's 'interpolation' remained hidden to both its author and his contemporaries. For this reason Planck felt compelled to search for a more rational foundation which he presented on December 14th, 1900, to a meeting of the Physical Society [in Berlin]. For the first time Planck accepts here without mental reservations Boltzmann's kinetic theory of heat according to which the entropy of a system in the [macro]state Z is given by

$$S(Z) = k \ln W(Z) \tag{28}$$

where $W(Z)$, called *the thermodynamic probability of Z*, is the number of microstates or 'complexions' realizing the macrostate Z. Quite obviously, this micro-statistical interpretation of entropy is something quite different from the previous conception of 'natural radiation'. Let us follow Planck's new approach which, as is well known, led to the *conscious* introduction of energy quanta $\varepsilon = h\nu$.

Again Planck considers a cavity filled with thermal radiation; however, instead of *one* oscillator, *very many* oscillators with *different frequencies* are now assumed to be present, N, say, having the frequency v, N' the frequency v', N'' the frequency v'', etc., where all N are large numbers. Let E_t be the total energy of the system (radiation + oscillators).

The question is how, in the stationary state, this energy is distributed among the resonators and the colours of the ... radiation, and what temperature the whole system then has.
 To start with, let

$$E_0 = E + E' + E'' + \dots$$

be the energy taken up by the oscillators with frequency v, v', v'', ..., respectively.
 Next, the energy taken up by all oscillators of the same frequency has to be distributed among these oscillators, e.g., the energy E among the N oscillators of frequency v. This can be done in an infinite number of ways if E is considered to be a quantity of unlimited divisibility. In contrast to this we consider – and this is the essential point in the whole calculation – E to be composed of a definite number of equal elements of finite magnitude and use for this purpose the natural constant $h = 6.55 \ 10^{-27}$ erg sec. This constant, multiplied by the frequency v common to the [N] oscillators, yields the energy element ε in erg, and when dividing E by ε we obtain the number P of energy elements that have to be distributed among the N resonators.

The number of different ways in which such a distribution can be effected, i.e.,

$$\frac{(N + P - 1)!}{(N - 1)! \, P!} \simeq \frac{(N + P)^{N+P}}{N^N P^P}, \tag{29}$$

is the number of 'complexions' W_v belonging to the frequency v (as far as the oscillators [or "resonators"] are concerned). The product of all W_v is the thermodynamic probability W [for the ensemble of oscillators]. One now has to find the values of the P for which W takes its maximum value under the condition $\Sigma Phv = \text{const.}$ (conservation of energy). This yields

$$P = \frac{N}{e^{\beta h v} - 1}, \tag{30}$$

Thus the decisive expression in Planck's radiation law is established. (We have left out of account the – entirely unnecessary – complication resulting from the fact that Planck still considered the oscillators as separate entities instead of identifying them with the [electromagnetic] eigenvibrations of the cavity).

 It will be seen from the above account that the introduction of discrete energy elements is a mere *consequence* of applying the statistical method of Boltzmann to radiation. As the hypothesis of energy quanta is already contained in Wien's radiation formula, *the novelty in Planck's new founda-*

tion is [not – as Planck thought – the introduction of radiation quanta but] *the application of Boltzmann's statistical method to radiation*. Hence the question arises as to the relation of this method to the method of averaging used previously in connection with the hypothesis of 'natural radiation'.

In the first place, it has to be emphasized again that the method previously used did not lead to a unique solution of the problem while the present method does. This shows clearly that the difference between the two methods is of a *fundamental* nature. Today, it is not difficult to see what this fundamental difference is.

There exists indeed, as we now know, a *natural line width* in [atomic] radiation, as may already be inferred from the limited coherence length of natural radiation. However, *this phenomenon has nothing to do with disorder or irregularity in the sense of the statistical theory of heat*. Physically, it is due to the reaction of emitted radiation on the emitter and its regular features are only understandable on the basis of quantum electrodynamics. True, a certain kind of statistical irregularity is implied by the natural line width as understood by quantum theory, but only in the same sense in which Heisenberg's indeterminacy relations have a statistical meaning: the natural line width [in frequency measure] Δv is a direct measure of the energy indeterminacy $\Delta E = h \Delta v$ of the emitted photon, and hence, in view of $\Delta E \, \Delta t \geqslant h$, a [reciprocal] measure of the indeterminacy of the time of emission, i.e., of the mean lifetime of the excited state of the emitting atom, Δt.[14] As the mean lifetime is inversely proportional to the emission probability [per unit time] the disappearance of the natural line width would imply an unlimited lifetime and hence the absence of any emission.

In view of these facts it is no longer surprising that the hypothesis of 'natural radiation' proved quite irrelevant to the solution of the problem. This resolves the 'paradox' mentioned above.

Strangely enough, this second foundation of Planck's law was likewise considered as unsatisfactory, indeed it was regarded as pure heresy. The first reason for this view was the unfamiliar way of determining the number of complexions which differed from the way used in gas statistics where it leads to the Maxwell-Boltzmann distribution. In spite of this, Planck's calculation is entirely in the spirit of Boltzmann's definition [of entropy]. Moreover, from the physical point of view it hardly differs from the calculating method later (1924) used by Bose, now known as *Bose-Einstein statistics*. Hence, the proper founder of this statistics was Planck, though Einstein was the first to discover its applicability to gas theory.

A further reason why Planck's computation was regarded as unintelligible seems to lie in its apparent inconsistency: on the one hand, discrete energy quanta are introduced, on the other, these quanta are not treated as Newtonian light particles or the particles of current gas theory in that quanta of equal frequency are taken to be *indistinguishable in principle*. To be sure, this is the very essence of Planck's mode of computing the number of complexions, and hence hardly a *new* reason for objecting. Still, the seeming inconsistency of the unfamiliar method was an additional aspect causing additional perplexity. Indeed, 'indistinguishable in principle' does not merely mean that the objects concerned are absolutely alike [such as mass-produced articles or twins from the same egg] but that their *hypothetical mutual exchange* [or *permutation*] *does not correspond to any real process whatsoever*. This becomes understandable if the objects concerned – viz., the light quanta of equal frequency – are *not regarded as independent entities but rather as excited states of a field*.

Thus, Planck's second foundation of his radiation formula has a much deeper physical significance than his first 'interpolation': while the 'interpolation' merely implies that the correct theory must be some sort of synthesis of wave and particle theory, the second foundation reveals a characteristic property of this synthesis, namely the *ontological priority of the field concept* [for radiation and indeed for all bosons].

Oddly enough, this implication of Planck's radiation law and its second foundation has not been generally recognized up to this very day. The reasons have their roots partly in the history of physics and partly in a mistaken philosophy. As for the first, it suffices to refer to the long period of the so-called classical quantum theory and the following period of quantum mechanics; both theories start from the corpuscular picture of matter. True, there have been attempts at interpreting quantum mechanics in terms of a classical wave theory ('matter waves') but such attempts were doomed to failure since they were based on a confusion of state function and wave function and the distinction between the two became obvious in the transition from the one to the many particle problem. Of longer life were the attempts at presenting quantum mechanics as a true synthesis of particle and wave theory. These attempts could draw support from two groups of facts. The first group consists of all those consequences of 1-particle quantum mechanics which on first sight seem to coincide with the results of classical wave theory. A typical example are the diffraction phenomena. The quantum mechanical treatment of the diffraction problem[15] confirms however the orthodox view according to which the diffraction pattern results from a great number of elementary processes of corpuscular character, each elementary process being connected with a stochastic exchange of momentum between particle and the experimental set-up [e.g., diffraction screen + photographic plate].

Of more interest [from the principal point of view] are those attempts that refer for support to the quantum mechanical treatment of the many body problem. In order to obtain correct results the state function, as is well known, has to be either symmetric (Bose-Einstein case) or antisymmetric (Fermi-Dirac case) in the coordinates of like particles. The 'statistics' resulting therefrom imply indeed, as outlined above, that the 'particles' belong to a common field. However, the decisive question is whether quantum mechanics merely *admits* or else *demands* that the state function be either symmetric or antisymmetric. Now the reasons advanced for the second answer all amount to saying that like particles are *practically indistinguishable*. This, however, is not sufficient [for establishing the non-Boltzmann types of 'statistics'] as the particles of classical gas theory are likewise practically indistinguishable! [Indeed, the argument quoted is typical of the positivist confusion of *meaning* and *testability*, or *semantics* and *pragmatics*]. Hence, the *use of the novel kinds of 'statistics' in the frame of quantum mechanics is to be regarded as a partial anticipation of quantum field theory* [where they can be *deduced*], in about the same sense in which Bohr's postulates and the entire "classical" quantum theory are partial anticipations of quantum mechanics: in all these cases the theoretical possibilities are restricted by additional postulates [that are foreign to the spirit of the theory concerned]. That these postulates contain the germ of a new theory merely shows the dialectics of historical evolution.

Another objection to Planck's [second] foundation was raised after one had learned how to quantize other systems, besides oscillators, by using either the rules of Bohr-Sommerfeld's 'classical' quantum theory or those of quantum mechanics. It then turned out that the oscillator is the only system where the energy increases in *equal* steps *hv*. (This is the second fortunate feature of Planck's choice mentioned above). Hence – so the objection – it is only due to this fortunate choice that Planck obtained correct results. The critics overlooked that this objection is only valid as long as the oscillators are interpreted as atomic models and that nothing is changed in Planck's conclusions if the oscillators are identified with the eigenvibrations of cavity-enclosed electromagnetic field.

I have deliberately put those objections into the foreground the discussion of which appears best suited to reveal the rational core of Planck's considerations and their significance for the ensuing development of physics. However, for the physical thinking at the beginning of the century other objections were more characteristic; they are the subject of the following discussion.

IX

To the leading men in theoretical physics in the period 1900 to 1924, and particularly in the statistical theory of heat, belonged above all H. A. Lorentz (1853–1928), P. Ehrenfest (1880–1933), and A. Einstein (1879–1955). Let us start with the attitude shown by Lorentz who was only 5 years older than Planck.

In view of the contradiction between the classical radiation law (Rayleigh, 1900) and the experiments, LORENTZ had re-examined the conditions for the validity of the equipartition theorem[16] of classical statistical mechanics used by RAYLEIGH[17a]. Since these conditions could be shown to be satisfied by all systems describable by Hamiltonian equations, and hence also by the [classical] oscillator, Lorentz had initially felt inclined to agree with the compromise suggested by Jeans as a way out of the dilemma. According to this compromise, both the Rayleigh law and the measurements were correct and the contradictions between them were merely due to the [supposed] fact that the measurements did not represent thermal equilibrium values because the equilibrium state would only be reached after an infinite time. This rather incredible view, already proved to be untenable both for theoretical reasons

(Rayleigh) and for experimental reasons (Lummer and Pringsheim), now appeared to gather unexpected support from the authority of Lorentz.

This [in brief outline] is the prehistory of a short but revealing note[17b] by LORENTZ from the same year 1908. In this note Lorentz recognizes the validity of the arguments against Jeans' view mentioned above and continues:

Wir werden uns vorstellen muessen, dass der Energieaustausch zwischen ponderabler Materie und Aether [read: radiation] zustande kommt durch Vermittlung der von Planck angenommenen Resonatoren oder aehnlicher Teilchen, die sich aus irgendeinem Grund der Anwendung der Gibbsschen Saetze [equipartition theorem] entziehen.

This should be read twice to comprehend its full meaning: For ordinary (ponderable) matter the equipartition theorem is valid, for radiation ('aether') we have instead the Planck distribution, and the contradiction between the two – which is strengthened rather than weakened by the mechanistic aether theory – is to be removed by the intervention of a third kind of matter.

A similar attitude had been shown by Lorentz in the fundamental questions of Einstein's Special Relativity: On the one hand the old theory of space and time was assumed to hold, on the other hand the Einstein equations were also admitted to be correct, and the discrepancies between the two were to be explained by some hypothetical mechanism.

This attitude clearly illustrates the unsolvable dilemma in which the old materialism was caught by the new discoveries. It also shows that mechanistic materialism was about to change from a progressive into a reactionary philosophy offering a patchwork of ad-hoc hypotheses where questions of principle demanded principal answers. At the same time the role of experiment as the final court of appeal, always recognized by Planck, was practically denied, though in a shy and backward way. To be just, LORENTZ himself did find a way out of the dilemma only two years later[18] by retiring to the unassailable position of Gibbs' theory (canonical ensemble) and deducing therefrom Planck's law.

Much less dogmatic and far more subtle was the response by Ehrenfest who at that time was just preparing his famous article for the *Enzyklopaedie der Mathematischen Wissenschaften* and who was therefore quite familiar with the dialectics of micro- and macro-physical considerations.

In his first paper [19] on the subject EHRENFEST expressed the view that Planck's distribution law does not necessarily contradict Boltzmann's way of determining the number of complexions as the latter could be made to yield any distribution law if in addition to the customary conditions (conservation of energy and number of particles) a third condition is imposed that contains a suitable function of the problem parameters. Planck's quantization of the oscillator energy, he suggests, may be regarded as such an additional condition. He does not, however, carry out this idea so that the question remained unsettled. [To be sure, auxiliary conditions imposed on an extremal or variational problem result in Lagrange parameters appearing in the solution of the problem; this should have made it clear that a reconciliation between the Planck law and 'Boltzmann statistics' in the way suggested by Ehrenfest was impossible.]

Five years later, after the Solvay Congress, EHRENFEST returns to the problem, this time approaching it from a different and more general point of view. [20] The central question is now whether Planck's quantization of the oscillator energy is *necessary* for obtaining Planck's distribution law. For reasons of maximum generality Ehrenfest introduces not only a *distribution function* depending on *both frequency* and *energy*, without assuming any connection between them, but also a *weight function* likewise depending on *both* parameters. In the first part of the paper he shows that the weight function has to satisfy certain conditions if the Boltzmann relation between entropy and thermodynamic probability is to be maintained. The Planck relation between frequency and energy then turns out to be a *necessary* condition for obtaining the Planck distribution.

Interesting and profound though this investigation is, it will appear as an unnecessary sophistication from the more intuitive point of view of Planck. Quite obviously, in Planck's opinion the number of complexions is uniquely determined by the given problem so that no place is left for introducing an arbitrary weight function or (as suggested by Einstein, cf. below) for a separate determination of *a priori* probabilities.

The difference between Planck's conception and that of Einstein is perhaps most clearly expressed by EINSTEIN himself in the introduction to his famous paper [21] (1909) in which he established a purely thermodynamical theory of fluctuation phenomena by inverting the Boltzmann-

Planck formula $S = k \ln W$ to obtain $W = \exp(S/k)$ [as a thermodynamical measure for the probability of states], and applied this theory to Planck's radiation law. There he writes:

Er (Planck) hätte das Gleichungspaar $S = (R/N) \lg W$, $W = $ Zahl der Komplexionen nur ansetzen dürfen, wenn er die Bedingung hinzugefügt hätte, daß die Komplexionen so gewählt werden müssen, daß sie in dem von ihm gewählten theoretischen Bilde *auf Grund statistischer Betrachtungen als gleichwahrscheinlich* befunden werden. Er wäre auf diesem Wege zu der von Jeans verteidigten (Rayleighschen, M. S.) Formel gelangt. (Italics by M.S.)

[In analysing this passage it will be seen that] Einstein does two [or rather three] things: [first, he *re-interprets* W to mean the number of *complexions having equal a priori probability*, secondly] he *demands* that this *equality be deduced from a theoretical picture by means of statistical considerations*, and finally he *reproaches* Planck for not having satisfied this demand. This *reproach* is in any case unjustified, no matter whether the *demand* is accepted or not. Indeed, if you break new ground in theoretical physics as Planck did with his radiation law you don't possess a complete 'theoretical picture' from which everything can be deduced, and the correct question in the sense of Einstein's demand would have been: *what complexions are supposed by Planck to have equal a priori probabilities, and what consequences or hints can be derived therefrom for the further development of the theory.*

As to Einstein's *demand* [which is presented as a sort of *methodological principle*], it must be said that, taken verbally, it is unrealizable and hence nonsensical. It is indeed a logical truism that a probability statement can be deduced, if at all, only from other probability statements by means of the calculus of probability, but not from any 'theoretical picture' and/or the calculus of probability ('statistical considerations') as such; even the equal probabilities for the faces of a die are founded on symmetry and not on statistical considerations. Hence, if Einstein's demand is to have any sense at all, the restriction "auf Grund statistischer Betrachtungen" has to be dropped. It then remains the demand that the equiprobability of the complexions used by Planck be justified by *dynamics* [i.e., ergodic theory]. Even this reduced demand is highly controversial [22], quite apart from the fact that it could not possibly be satisfied without first establishing the full-blown dynamical theory (viz., quantum electrodynamics). Since, moreover, any probabilistic hypothesis is a synthetic statement to be judged by its testable consequences it must be conceded that Planck was entirely right in paying no attention to the demands of ergodic theory. [Last not least, Einstein's re-interpretation of W, involving as it does the doubtful conception of *a priori* probabilities, can hardly be reconciled with the basic conception of the Boltzmann-Planck definition of entropy.]

X

The fact that Planck's work was received rather badly by contemporary theorists, notably Einstein, had the effect that the new kind of 'statistics' due to Planck was not named after him but received the name 'Bose-Einstein statistics'. Let us therefore consider BOSE's paper[23] which in a postscript by Einstein is termed 'an important advance'.

Bose starts with two objections. First, the factor $N = 8\pi v^2/c^3$ in Planck's law is taken over from the classical theory, and, second, Wien's displacement law, likewise used by Planck, rests also on classical electrodynamics.

A first reply to this is obvious: you cannot construct a new theory without taking over elements of the old theory; it is just the job of the theorist to distinguish between what can be maintained and what has to be changed.

In the present case the situation is as follows. The expression for N belongs to wave kinematics and could only be changed if a new universal constant of dimension 'length' or 'time' is introduced.

Somewhat more complicated is the situation with respect to Wien's displacement law. As a detailed analysis shows[24], this law rest – apart from the laws of thermodynamics – on precisely two independent conditions concerning radiation, one belonging to kinematics, the other to dynamics. The first condition reads

$$\frac{\delta v}{v} = -\frac{1}{3}\frac{\delta V}{V}$$

and connects the change of eigenfrequency of a reflecting cavity with the change in volume under adiabatic and isentropic [quasistatic] expansion of the cavity. The second condition reads

$$p(v) = \tfrac{1}{3}u(v)$$

and connects the spectral energy density $u(v)$ with the spectral partial pressure $p(v)$. To the first condition applies what has been said above about the equation $N = 8\pi v^2/c^3$. The maintenance of the second condition cannot be justified in this way [since the factor $\frac{1}{3}$ could be replaced by a function of the numerical variable $x = hv/kT$]. However, the second condition merely expresses the fact that there is *no [dynamical] interaction between the different modes of eigenvibrations or photons of the same frequency*, i.e., that the Maxwell equations are linear or that the photon gas is an 'ideal' gas, respectively.

As all this was, or could be, known in 1924, Bose's objections were untenable even then.

The 'novelty' in Bose's paper, as compared to the work of Planck, consists in the use of the *phase space* the re-introduction of which was a step back rather than a step forward: in quantum mechanics the [classical] phase space is replaced by a Hilbert space which is a space of *states*.

Incidentally, Bose himself never claimed to have done anything fundamentally new in statistical theory; indeed he writes:

Dagegen scheint mir die Lichtquantenhypothese in Verbindung mit der statistischen Mechanik (wie sie durch Planck den Bedürfnissen der Quantentheorie angepaßt worden ist), für die Ableitung des Gesetzes unabhängig von der klassischen Theorie hinreichend zu sein.

This, I hope, will end a widespread legend.

XI

While in his attitude to physical statistics Planck proved superior to his critics, the opposite is true for his attitude to Maxwell's theory. It is indeed one of the most remarkable inconsistencies in Planck's thinking that he was not prepared to draw the obvious consequences concerning the status of this theory. However, who is not prepared to advance will be compelled by the logic of circumstances to retreat.

The retreat started 1911–1912 with the re-introduction of continuous absorption and ended in 1914 with the re-introduction of continuous emission. While the 1911 and 1912 papers[25] still contained some very remarkable and progressive ideas, the 1914 paper[26] amounted to a complete abandonment of the very principles of natural science. Since neither the one nor the other aspect has been properly apprehended so far[27], an examination of these papers will now be given.

In the first of these papers (1911) Planck justifies the re-introduction of continuous absorption by referring to the well-known difficulty concerning the time required [by a classical atom] for absorbing a high energy quantum, a difficulty unresolvable in classical electrodynamics and not resolved but merely shifted to another place in the (classical) corpuscular theory of light revived by Einstein.

To be sure, the difficulty mentioned has no direct connection with the problem solved by Planck's distribution law, and Planck would have been completely justified to take the view that any possible deduction of his law from a theory of interaction between radiation and atoms must be preceeded by a corresponding development of atomic theory. But such a logical ordering of problems and a corresponding research strategy were not within the grasp of either Planck or his contemporaries; they were practiced later with remarkable success by Niels Bohr. In particular, Planck was hampered in this respect by still considering his 'resonators' as independent entities instead of identifying them with the eigenvibrations of cavity-enclosed radiation field. On the other hand, it would be unjust to reproach Planck for not having solved a problem that could not be solved at that time. One even has to admit that, from the present point of view, the re-introduction of continuous emission and absorption, taken by itself, contained a progressive element if compared to Einstein's corpuscular light theory and Bohr's theory of 1913. While these theories offer no understanding at all of the existence of a natural line width and a finite coherence length [28], Planck's conception [of continuous absorption and emission] anticipates to a certain degree the present solution of the problem, in so far, namely, as the present theory does not ascribe a sharp instant of time to the process of emission or absorption.[28]

Planck's 1911 paper contains another most remarkable anticipation of topical interest. Planck draws attention to the fact that the *lattice vibrations of a solid* and their quantum properties are decisive not only for the theory of the specific heat of solids (Einstein, 1907) but also for the *theory of thermal conductivity* [in solids]. Hence it may be expected, so Planck points out, that also in the energy exchange between the lattice vibrations the emitted energy is quantized. This is a true anticipation of the present *phonon theory of thermal conduction* [in solids] which has not only been confirmed by recent measurements but has also proved of practical importance for investigating solid structures.[29]

Finally, the same paper (1911) contains the first suggestion that *radio-activity* may be regarded as some kind of 'quantum emission' – a suggestion that has proved essentially correct. To this may be added that the paper anticipates the quantum mechanical value $h/2$ for the *zero point energy of the oscillator*.

In contrast to the 1911 paper, the following paper (1912) leaves the reader disappointed. Here, Planck explicates his idea of continuous absorption by computing, for his oscillator model, the 'mean time of accumulation' τ_1, i.e., 'the time required by the oscillator for absorbing one energy element from the surrounding black radiation'. He obtains the result [30]

$$\tau_1 = \frac{mc^2}{(e^2/\lambda)}\, v^{-1} \left(e^{hv/kT} - 1\right).$$

Theoretically this formula makes no sense because the energy expression (e^2/λ) in the denominator does not admit of a physical interpretation consistent with the model. [31]

Thus, this paper merely amounts to the – quite unnecessary – explication that a reconciliation of quantum theory and classical electrodynamics is impossible. Planck, to be sure, did not realize this and thus it remained to others to draw the consequences.

The 1914 paper [26], where also the quantum nature of *emission* is given up, derives its main interest for us from the fact that it illustrates a general lesson: any attempt at resolving the (quite natural and unavoidable) conceptual difficulties of a new theory within the conceptual framework of the old theories is bound to lead straight into metaphysics and irrationalism. Indeed, Planck writes:

Ich habe mir daher die Frage vorgelegt, ob man nicht noch einen Schritt weiter gehen kann und außer der Absorption auch die Emission der strahlenden Wärme als stetig voraussetzen kann. Diese Auffassung würde darauf hinauskommen, daß *die Quantenwirkungen gar nicht zwischen den Oszillatoren und der Wellenstrahlung, sondern lediglich zwischen den Oszillatoren und den freien Partikeln* (Molekülen, Ionen, Elektronen) *stattfinden*, welche bei den Zusammenstößen mit den Oszillatoren Energie austauschen. [Italics by M.S.]

Thus, Planck's oscillators, which up to now could be identified with the electromagnetic eigenvibrations of the cavity (in spite of Planck's different interpretation), are now being opposed as seperate entities not only to the radiation but also to the ordinary ponderable matter ('free particles'). This is precisely the same jump into metaphysics as the one suggested

three years before by Lorentz as a 'way out' from the contradiction between old theory and new experience: instead of regarding the latter as
the result of *newly discovered properties* of matter and hence modifying
the old theory, a novel kind of matter is invented not subjected to the
ordinary laws of nature – a veritable *deus ex machina* that ensures a
Happy End.

Still, Planck would not have been the physicist of genius he was, had
he not discovered even in this impasse some trace of a hidden law.

The first question Planck had to solve in carrying out his 'new' program
concerned the exchange of energy between the 'free particles' and the
mysterious oscillators fixed in space. According to the good old mechanistic view such an exchange of energy was to occur only in collision.
What, however, is to be understood by 'collision' in this context?
Planck defines:

Wenn eine Partikel mit der Geschwindigkeit q in der Nähe des Oszillators vorbeifliegt,
so soll ein Zusammenstoß nur und immer dann erfolgen, wenn das Produkt der Geschwindigkeit q und des Minimalabstandes r des Oszillatormittelpunktes von der gradlinigen
Bahn der herankommenden punktförmig gedachten Partikel kleiner ist als eine gewisse
Konstante

$$q \cdot r < f.$$

He further remarks that the quantity mf (m = mass of a particle) has the
dimension of action. Thus, if Planck had used the momentum $p = mq$
instead of the velocity q he could have written his defining condition in
the form

$$pr < ah \quad (a = \text{real number}),$$

This, to be sure, is not Heisenberg's indeterminacy relation, but it is a
definite step towards it. It says that the transition from $pr > ah$ to $pr < ah$
results in the appearance of a new quality (exchange of energy) and
further that all cases with $pr < ah$ are *indistinguishable* in this respect.

Even closer is the connection with the quantum mechanical scattering
theory since h/p is the De Broglie wavelength of the particle considered
and its ratio (though not to the distance r but) to the diameter of the
scatterer largely determines the character of the scattering.

In the same year (1914) when Planck tried to save Maxwell's theory
from the attack of quantum theory, a paper by A. D. Fokker [32] appeared
in which the equilibrium distribution of a system of *rigid rotating dipoles*
in a given radiation field according to Maxwell's theory is exactly com-

puted. The resulting expression for the rotational part of the specific heat of diatomic molecules differed from the experimental results both quantitatively and qualitatively. This induced PLANCK[33] to return to the quantum emission hypothesis and to establish equations for the interaction of radiation and ponderable matter the quantum character of which is quite obvious: instead of Fokker's differential equation

$$W(q)\,\bar{r} - \frac{1}{2}\frac{\mathrm{d}}{\mathrm{d}q}\left[W(q)\,\overline{r^2}\right] = W(q)\,\Delta q$$

Planck obtains the difference equation

$$\tfrac{1}{2}\overline{r_n^2}\left[W_n(q_n) - W_{n+1}(q_n)\right] = \Delta q_n \int_{q_n}^{q_{n+1}} W_{n+1}(q)\,\mathrm{d}q$$

where q is the [classical] state parameter [i.e., a point in phase space] and n the quantum number of the elementary cell in the [quantized] phase space. (As is easily seen, Planck's equation implies that an emission takes place only if the state parameter q falls into the interface between two adjacent elementary cells.)

Here, too, quantization is again applied to ponderable matter only, but Planck is now admitting that electrodynamics may also be in need of change:

Die naehere elektrodynamische Analyse der Gleichung bildet ein Problem, dessen weitere Verfolgung tiefer in die Erkenntnis des Wesens der Quantenhypothese einzufuehren verspricht. Doch moechte ein Versuch in dieser Richtung hier noch verfrueht erscheinen.

Indeed, quantum mechanics had first to develop the mathematical methods by means of which the quantization of the electromagnetic field could be effected.

After quantum electrodynamics had been established [1927] Planck took no notice of it, neither in his writings nor in his lectures. Up to the last years of his life he wrestled with the problems of nonrelativistic quantum mechanics without being able to master them, as may be seen from his last three papers[34] which provide a moving demonstration of this fact.

XII

If, in conclusion, we consider the present state of quantum theory the first thing to be stressed is the further unification of our physical world picture which, it will be remembered, was the supreme aim of all Planck's endeavours. Indeed, the quantum theory of fields not only explains the empirical relation between spin value and 'statistics' [permutational parity], unexplained by quantum mechanics, but it is the first theory that accounts for both the particle and the wave properties of matter in a natural way. Of even greater importance from the philosophical point of view is the fact that the quantum field theory has overcome the old metaphysical contrast between 'matter' and 'force' in a truly dialectical manner. While classical mechanics and quantum mechanics account for the action of forces by separate laws of forces or potentials which have no logical connection with the other axioms and hence do not represent an inherent property of matter, quantum field theory [reduces forces to the (virtual) exchange of field quanta and hence] makes forcelike action an inseparable aspect of matter.

On the other hand, the well-known internal difficulties of quantum field theory as well as its incapability to explain the empirical mass spectrum and the empirical values of the coupling constants clearly point to a necessary and radical change of fundamental theory which probably will be connected with the introduction of a further universal constant having the physical dimension of 'length'. This new theory will certainly not mark a return to the outdated ideas of mechanical material-ism but will represent a dialectical negation or uplation of the present quantum field theory. Whether Planck's radiation law will be modified by the new theory cannot be decided at present, though the theoretical possibility of such a modification has been definitely established[35].

Finally, the significance of Planck's law for the evolution of physics does not lie so much in the accidental fact that quantum theory originated from it – quantum theory could have equally well started from the photoelectric effect, from the anomalies of the specific heats, from the empirical laws of the line spectra, or from other phenomena; rather, its significance lies in the fact that the law itself as well as the two foundations provided by Planck are a signpost leading directly into the quantum field theory.

Failure to see this makes the tragedy in Planck's scientific life, just as it mars the arguments of his critics.

BIBLIOGRAPHY – A HISTORICAL SURVEY

The following bibliography provides a survey on the historical development of the subject, including the prehistory of Planck's work. The grouping under different headings serves to emphasize the main lines of development. *Statistical Thermodynamics* has not been included.

The parenthesized number behind any of the Planck papers quoted in the following is the ordinal number under which that paper is reprinted in *Max Planck, Physikalische Abhandlungen und Vortraege* (3 volumes), Braunschweig 1958.

1. THERMODYNAMICS

S. Carnot	1824	*Réflexions sur la Puissance Motrice du Feu.* Paris 1824.
(1796–1832)		(Maximum efficiency of heat engines).
W. Thomson	1848	*Proc. Camb. Phil. Soc.* **1**, 66.
(Lord Kelvin)		*Phil. Mag.* **33**, 313.
(1824–1907)	1851	*Trans. R. S. Edin.* **20**, 261.
R. Clausius	1850	'Über die bewegende Kraft der Wärme', *Ann. d. Phys.* **79**, 368, 500.
(1822–1888)		
	1865	'Über verschiedene für die Anwendung bequeme Formen der Hauptgleichungen der mechanischen Wärmetheorie', *Ann. d. Phys.* **125**, 353.
		(Absolute temperature and Second Law.)
M. Planck	1879	*Über den zweiten Hauptsatz der mechanischen Wärmetheorie.* Inauguraldissertation, München. (1).
(1858–1947)		
	1887 –1891	'Über das Prinzip der Vermehrung der Entropie. 1. bis 4. Abhandlung.' *Ann. d. Phys.* **30**, 562; **31**, 189; **32**, 462; **44**, 385. (7), (8), (9), (19).
		(General formulation of Second Law and applications.)

2. THERMODYNAMICAL THEORY OF RADIATION

P. Prévost	1792	*Recherches physico-mechaniques sur la chaleur.* Genève.
(1751–1839)	1809	*Essai sur la calorique rayonnant.* Genève.
		(Theory of thermal equilibrium by exchange of radiant energy.)

(Prévost's idea that even after the attainment of thermal equilibrium the exchange of thermal radiation does not cease represents one of the most magnificent achievements of creative imagination and dialectical thinking in his time. This idea is the very basis of the entire theory of thermal radiation and the prototype of all following dynamical theories of equilibrium states, including chemical ones.)

B. Stewart	1858	Trans. R. S. Edin. **22**, 1.
(1828–1887)		
G. Kirchhoff	1859	*Berl. Monatsber. 1859*, 622, 783.
(1824–1887)	1860	*Ann. d. Phys.* **109**, 148, 275.
	1860	*Berl. Abhandl.* (Phys.), 63.

1862   *Berl. Abhandl.* (Phys.), 227.

1863   *Ann. d. Phys.* **118**, 94.
('Kirchhoff's Law': proportionality between spectral emissivity and spectral absorptivity; hence: existence of a universal spectral energy distribution function for thermal ('blackbody') radiation.)

J. Stefan   1879   *Wien. Ber.* **79** (Abt. II), 391.
(1835–1893)   (T^4-law for integral thermal radiation experimentally established.)

M. Strauss   1961   'On the Possible Generalizations of Planck's Law', *Nuov.*
(1907–   )   *Cim.* (X) 19, 594–596.

1962   'Verallgemeinerung des Planckschen Strahlungsgesetzes in *h-c-l*-Theorien', *Z. Naturf.* 17a, 827–847.
(Contains a partial differential equation for the spectral distribution function which, together with two conditions of thermodynamic consistency, represents the *maximum statement on the distribution function derivable from thermodynamics*; this answers the question put by Planck in his 1900 paper referred to in note 11a. From the differential equation mentioned the *necessary and sufficient conditions for the validity of Wien's displacement law* are derived.)

3. MAXWELL'S ELECTRODYNAMICS

J. C. Maxwell   1873   *Treatise on Electricity and Magnetism*
(1831–1879)   (§ 792 gives a quantitative treatment of *light pressure*.)

H. Hertz   1889   *Ann. d. Phys.* **36**, 12 (Hertzian oscillator).
(1857–1894)

M. Planck   1896   'Absorption und Emission elektrischer Wellen durch Reso-
(1858–1947)   nanz', *Ann. d. Phys.* **57**, 1–14 (26).

1897   'Über elektrische Schwingungen, welche durch Resonanz erregt und durch Strahlung gedämpft werden', *Ann. d. Phys.* **60**, 577–599. (28).

1897   'Notiz zur Theorie der Dämpfung elektrischer Schwingungen', *Ann. d. Phys.* **63**, 419–422 (29).

4. THEORY OF THERMAL RADIATION, WITH USE OF ELECTRODYNAMICS

L. Boltzmann   1884   Ann. d. Phys. **22**, 31, 291.
(1844–1906)   (Theoretical proof of Stefan's T^4-law.)

W. Wien   1893   *Ber. Preuss. Akad. Wiss. 1893*, 55.
(1864–1924)   1894   *Ann. d. Phys.* **52**, 132.
(Wien's displacement law, cf. note 6.)

M. Planck   1897   'Über irreversible Strahlungsvorgänge. 1. bis 5. Mittlg',
(1858–1947)   –1899   *Ber. Preuss. Akad. Wiss.* (30)–(34).

1900   'Über irreversible Strahlungsvorgänge', *Ann. d. Phys.* (4) **1**, 69–122. (36).

Lord Rayleigh   1900   *Phil. Mag.* **49**, 539.
(1842–1919)   1905   *Nature* **72**, 54, 243.
(Radiation law corresponding to Maxwell's theory [and the equipartition theorem].)

5. PLANCK'S LAW

M. Planck   1900   'Über eine Verbesserung der Wienschen Spektralgleichung',
                              Verh. Dt. Phys. Ges. **2**, 202–204 (38).
                              ('Interpolation', cf. text.)

                       1900   'Zur Theorie des Gesetzes der Energieverteilung im Normal-
                              spektrum', *Verh. Dt. Phys. Ges.* **2**, 237–245 (41).

                       1901   'Über das Gesetz der Energieverteilung im Normalspektrum',
                              Ann. d. Phys. (4) **4**, 553–563 (43).
                              (Quantization of oscillator energy, quantum statistics, cf.
                              text.)

H. A. Lorentz   1910   *Phys. Zs.* **11**, 1234.
(1853–1928)   (Use of Gibbs' canonical ensemble for obtaining mean oscil-
                              lator energy.)

A. Einstein   1917   'Zur Quantentheorie der Strahlung', *Phys. Zs.* **18**, 121.
(1879–1955)   (Derivation of Planck's law from dynamical equilibrium
                              between radiation and Bohr atoms.)

6. FURTHER WORKS

A. Einstein   1909   'Zum gegenwärtigen Stand des Strahlungsproblems', *Phys.
                              Zs.* **10**, 185.
                              (Thermodynamical theory of energy fluctuations applied to
                              Planck radiation.)

M. Planck   1911   'Eine neue Strahlungshypothese', *Verh. d. Dt. Phys. Ges.* **13**,
                              138–148 (73).
                              'Zur Hypothese der Quantenemission', *Sitz. Ber. Preuss.
                              Akad. Wiss.* (1911), 723–731 (74).

                       1912   'Über die Begründung des Gesetzes der schwarzen Strahlung',
                              Ann. d. Phys. (4) **37**, 642–656 (76).
                              (Continuous absorption, cf. text.)

                       1911   'Die Gesetze der Wärmestrahlung und die Hypothese der
                              elementaren Wirkungsquanten', *Solvay Congress*, Brussels
                              1911 (75).

                       1914   'Eine veränderte Formulierung der Quantenhypothese', *Sitz.
                              Ber. Preuss. Akad. Wiss.* (1914) 918–923 (79).
                              (Continuous absorption and emission, cf. text.)

                       1915   'Über Quantenwirkungen in der Elektrodynamik', *Sitz. Ber.
                              Preuss. Akad. Wiss.* (1915) 512–519 (81).
                              (Quantum emission re-introduced, cf. text.)

P. Debye   1910   *Ann. d. Phys.* **33**, 1427 (Cf. also:)
A. Rubinowicz   1917   *Phys. Zs.* **18**, 96.
F. Hasenoehrl   1911   *Phys. Zs.* **12**, 931 (Gibbs' ensemble.)
Ph. Frank   1912   *Phys. Zs.* **13**, 506 (Proof that the oscillators of Planck's 1911
                              theory form a canonical ensemble.)

A. Einstein   1923   (and P. Ehrenfest) 'Zur Quantentheorie des Strahlungs-
                              gleichgewichts', *Z. Phys.* **19**, 301.

S. N. Bose   1924   'Plancks Gesetz und die Lichtquanten', *Z. Phys.* **26**, 178–181
                              (Cf. text).

K. F. Novobatzky   1958   'Strahlungs- und Gasstatistik' in *Max-Planck-Festschrift*,
                              Berlin 1958.

NOTES

* Translated from 'Max Planck und die Entstehung der Quantentheorie' in *Forschen und Wirken* – Festschrift zur 150-Jahr-Feier der Humboldt-Universitaet zu Berlin, Bd. I, Berlin 1960.

[0] [Cf. note 22].

[1] M. Planck, *Wissenschaftliche Selbstbiographie*, Leipzig 1948, p. 11.

[2] Ibid., p. 12. [Translation by the author.]

[3] Gibbs' thermodynamical papers, of which the third ('On the equilibrium of heterogeneous substances') became the most famous, appeared 1873 in the little read *Trans. of the Connecticut Acad.* and remained widely unknown for a rather long time although C. Maxwell had immediately pointed out their importance. An interesting study of Gibbs and his relation to society has been given by J. G. Crowther, *Famous American Men of Science*, vol. II, Pelican Books, London 1937.

[4] *Ann. d. Phys.* **19** (1883) 358.

[5] *Ann. d. Phys.* **15** (1882) 446.

[6] By Wien's displacement law we mean the equation

$$u(v, T) = v^3 f(v/T), \tag{a}$$

where the left-hand side represents the spectral energy density at the temperature T and where f is an unknown function. The significance of this law consists in the fact that it reduces an unknown function of two variables to an unknown function of one variable (v/T).

If the spectral energy distribution at given temperature has exactly one maximum, as confirmed by experiments, eq. (a) implies

$$v_{max}/T = v'_{max}/T' \quad \text{or} \quad \lambda_{max}T = \lambda'_{max}T'. \tag{b}$$

Often (particular by experimental physicists) this latter equation is termed 'Wien's displacement law'.

[Equation (a) is the general solution of, and hence equivalent to, the partial differential equation

$$vu_v + Tu_T = 3u \tag{c}$$

which is a special case of the 'fundamental equation' established by the author in 1960, cf. note 24.]

[7] The theory of physical dimensions is due to Fourier who explains it in his famous work *Théorie Analytique de la Chaleur* (1822) (§ 157). Later, dimensional analysis was applied with great success particularly by Lord Rayleigh to many problems (e.g., blue colour of the sky by scattering of sunlight).

[8] The only expression having the same physical dimension as u that can be formed from c, kT, and v is $v^2c^{-3}kT$ so that

$$u(v, T) = nv^2c^{-3}kT.$$

This is indeed Rayleigh's radiation law, apart from the unknown numerical factor n. According to this law the spectral energy density increases without limit with increasing frequency v so that the integrated energy density becomes infinite ('ultraviolet catastrophe'). This suffices to conclude that Maxwell's theory [is incompatible with thermodynamics and hence] has to be modified so as to contain a further universal constant (h).

Furthermore, since Rayleigh's law follows from the equipartition theorem of classical statistical mechanics, it must be concluded that classical mechanics, too, is to be changed in such a way that the new mechanics contains likewise the new constant (h).

[9] For the following cf. the papers 'Ueber irreversible Strahlungsvorgaenge' (1897–1900), in particular the last of these papers, *Ann. d. Phys.* (4) **1** (1900) 69–122.

[10] Personal reminiscence of the author.

[11] 'Entropie und Temperatur strahlender Waerme', *Ann. d. Phys.* (4) **1** (1900) 719–737.

[11a] Ibid.

[12] 'Zur Theorie der Waermestrahlung', *Ann. d. Phys.* (4) **31** (1910), 758–768.

[13] *Verh. Dtsch. Phys. Ges.* (3) **19** (1938), 62.

[14] The coherence length is then given by

$$\Delta l = c\Delta t = c/\Delta v.$$

Experimentally, Δl is of the order of magnitude of 1 m, so that the mean lifetime Δt of an excited state becomes about 10^{-8} s which agrees with experimental values. Since according to quantum theory a photon can [normally] only interfere with itself, the coherence length measures the mean extension of a photon.

[15] Paper by M. Strauss, read at the Meeting of the Physikalische Gesellschaft der DDR in Halle, 1954. (Summary in *Bericht ueber die Arbeitstagung 'Angewandte Physik' vom 3. bis 4. Sept. 1954 in Halle/S*, Leipzig 1955.)

[16] The equipartition theorem says that in thermal equilibrium each (simple) degree of freedom takes up the (mean) energy $kT/2$; (the degree of freedom of the linear oscillator has to be counted double since the oscillator takes up kinetic as well as potential energy; similarly, the energy of electromagnetic oscillations is half electric and half magnetic). – Cf. also note 8.

[17a] H. A. Lorentz, *Nuovo Cim.* **16** (1908), 5.

[17b] H. A. Lorentz, *Phys. Zs.* **9** (1908), 562.

[18] H. A. Lorentz, *Phys. Zs.* **11** (1910), 1234.

[19] P. Ehrenfest, *Phys. Zs.* **7** (1906), 528–532.

[20] P. Ehrenfest, 'Welche Zuege der Lichtquantenhypothese spielen in der Theorie der Waermestrahlung eine wesentliche Rolle', *Ann. d. Phys.* **36** (1911), 91–118.

[21] A. Einstein, 'Zum gegenwaertigen Stand des Strahlungsproblems', *Phys. Zs.* **10** (1909), 185–193.

[22] The strife about the foundations of statistical thermodynamics is essentially the old controversy between the followers of Gibbs and the proponents of ergodic theory, often called the continental school. Ergodic theory aims at a dynamical foundation of statistical thermodynamics by proving the equation: statistical mean value [in a Gibbs ensemble] = mean value [in phase or state space] taken over infinite time. Since ergodic theory thus depends on the dynamical theory just available while the theorems of statistical thermodynamics claim general validity it appears doubtful whether the aim of ergodic theory is at all sensible. [For more detailed objections cf. E. T. Jaynes, 'Foundations of Probability Theory and Statistical Mechanics' in *Delaware Seminar in the Foundations of Physics* (ed. by M. Bunge), Berlin Heidelberg New York 1967.] The defenders of ergodic theory are committed to the view that a dynamical theory not satisfying the above equation is defective in principle. This view, which I regard as rather dogmatic if not metaphysical, is also at the basic of the interesting studies by L. Rosenfeld ('La première phase de l'évolution de la Théorie des Quanta', *Osiris* 2 (1936), 149, and particularly 'Max Planck et la définition statistique de l'entropie', in *Max-Planck-Festschrift*, Berlin 1958. [Cf. also 'On the Foundations of Statistical Thermodynamics', *Acta Physica Polonica* 14 (1955), 3–39]). If in these papers Gibbs and Planck are charged with holding idealist views in questions

of statistical thermodynamics it seems necessary to reply that scientific abstraction should not be confused with idealism.

[23] N. Bose, 'Plancks Gesetz und die Lichtquantenhypothese', *Z. Phys.* **26** (1924), 178–181.

[24] M. Strauss, 'Verallgemeinerung des Planckschen Gesetzes bei Beruecksichtigung der Existenz einer dritten universellen Naturkonstante', *Monatsber. Dtsch. Akad. Wiss. Berlin* **2** (1960), 412–417; ['On the Possible Generalizations of Planck's Law', *Nuov. Cim.* (X) **19** (1961), 594–569; 'Verallgemeinerung des Planckschen Strahlungsgesetzes' in *h-c-l-*Theorien', *Z. Naturforschung* **17a** (1962), 827–847.]

[25] M. Planck, 'Eine neue Strahlungshypothese', *Verh. d. Deutsch. Phys. Ges.* **13** (1911), 138–148. 'Ueber die Begruendung des Gesetzes der schwarzen Strahlung', *Ann. d. Phys.* (4) **37** (1912), 642–656.

[26] M. Planck, 'Eine veraenderte Formulierung der Quantenhypothese', *S.-B. Preuss. Akad. Wiss.* 1914, 918–923.

[27] Cf., e.g., L. Rosenfeld's paper referred to in note 22 and E. Whittaker, *History of the Theories of Aether and Electricity*, vol. II, London 1953, pp. 103–104.

[28] Cf. note 14. – Later (1923) Planck himself drew attention to the relation between coherence length and 'Abklingungszeit' (i.e., mean lifetime of an excited state) and to the difficulty it presents to Bohr's theory; cf. M. Planck, 'Die Bohrsche Atomtheorie', *Naturwiss.* **11** (1923), 535–537.

[29] K. Mendelssohn, lecture given to the *Max-von-Laue-Kolloquium* Berlin, October 2, 1959.

[30] This equation follows from the formulae (28), (26), (32) of Planck's 1912 paper quoted in note 25.

[31] The expression e^2/λ represents the electrostatic energy of two elementary charges separated by a distance of one *wavelength* (!). If instead of the oscillator mass m the amplitude q_n of the quantized oscillator is introduced via

$$(1/2m)\left(p^2 + m^2\omega^2 q^2\right) = \tfrac{1}{2}m\omega^2 q_n^2 = \left(\tfrac{1}{2} + n\right)h\nu$$

one obtains for the first factor in the expression for τ_1:

$$\frac{mc^2}{e^2/\lambda} = (2n + 1)\frac{hc}{e^2}\,(\lambda/q_n)^2\,(\nu/\omega)^2\,;$$

this expression, though it contains the fine structure constant in the correct way, does not make physical sense (ω is the eigenfrequency of the oscillator). – That the formula given in the text is theoretically unsound may also be seen from the fact that it is not Lorentz covariant.

[32] A. D. Fokker, 'Die mittlere Energie rotierender Dipole im Strahlungsfeld', *Ann. d. Phys.* (4) **43** (1914), 810–820.

[33] M. Planck, 'Ueber Quantenwirkungen in der Elektrodynamik', *S.-B. Preuss. Akad. Wiss.* 1915, 512–519.

[34] M. Planck, 'Versuch einer Synthese zwischen Wellenmechanik und Korpuskularmechanik', *Ann. d. Phys.* (5) **37** (1940), 261; **38** (1940), 272; **40** (1941), 481.

[35] Cf. note 24.

PART B

LOGIC OF PHYSICS – GENERAL

CHAPTER V

ON THE RELATION
BETWEEN MATHEMATICS AND PHYSICS AND
ITS HISTORICAL DEVELOPMENT*

In January, 1966, some mathematicians and physicists, invited by the Secretary of the *Fachbereich Physik* of the *Forschungsgemeinschaft* [of the Academy], met to exchange their views on the relation between the disciplines they represented. The discussion was based on the 'Thesen zum gegenwaertigen Verhaeltnis zwischen Mathematik und Physik' ('Theses on the Present Relation between Mathematics and Physics') which have been published in the meantime.[1]

The following remarks are a revised and enlarged version of the remarks made by the author on that occasion.

0. The alienation between Mathematics and Physics, noted and deplored in the 'Theses', is without doubt the result of a historical development; however, the question to be asked is this: how far is this result a necessary consequence of developmental laws and how far does it depend on factors that can be socially controlled? A further question to be asked is this: will this development continue in the same direction, or is there an objective possibility of overcoming the said alienation?

These questions belong to the science of science; their solution requires a logico-historical analysis of the interrelations between Mathematics and Physics.

1. There are good reasons for calling Physics the mother of Mathematics. It is not correct, however, to say that *all* parts of mathematics owe their origin to physics. Thus, the *classical theory of probability* (Pascal, Fermat, J. Bernoulli) originated as the mathematical theory of games of chance, the *theory of games* (von Neumann) as the theory of optimal strategy in conflict situations, *information theory* originated from the requirements of communication engineering, and *cybernetics* from those of automation. Moreover, *modern algebra* owes much to logic (Boole).

These and other examples show also that the resulting mathematical theories, thanks to their high level of abstraction, have a range of application immensely larger than the original domain. Thus, *the origin of a mathematical theory is not at all characteristic of the theory; from the viewpoint of mathematics it is immaterial and accidental. This simple fact suffices to exclude any naturalistic conception of mathematics.*

2. While up to the end of the 18th century the development of mathematics was dominated almost exclusively by external demands, above all by those of physics, the first half of the 19th century marks the beginning of the *emancipation* and *self-determination* of mathematics (Cauchy, Abel, Galois, Jacobi) which we find fully developed in the second half of that century, e.g. in the work of Weierstrass (1815–1898) and Dedekind (1831–1918). The aspiration for a mathematics independent of empirical foundation led – after technically and philosophically insufficient attempts by Kronecker ('arithmetization') and Cantor (reduction to set theory) – with inherent necessity to the [modern] view that all *mathematical 'theories' are calculi that are subjected to the condition of internal consistency only.*

This view found its classical expression in Hilbert's metamathematical *proof theory*. According to it the proper content of all mathematics consists in statements of the form: formula *A* is *deducible* from the primary formulae (axioms) according to logico-mathematical rules of deduction [or transformation]. Hence it makes no sense to ask whether these axioms are *true* (i.e., agreeing with reality) or *evident*; such questions are meaningful only for physically or otherwise *interpreted* calculi.

The same 'formalistic' conception of mathematics is shared by EINSTEIN[2]:

. "In so far as the statements of mathematics refer to reality they are insecure, in so far as they are secure they don't refer to reality"[3]. In a similar way Lichtenberg had defined mathematics as *Wissenschaft von dem, was sich von selbst versteht* [thus expressing the analytic character of it].

3. By its emancipation from physics and other sources mathematics gained the freedom to develop according to its own laws. Thereby that *advance of mathematics beyond* [the given requirements of] *physics* became possible without which modern physics would be non-existent. I only refer to the *tensor calculus* (Ricci, Levi-Civita) as the mathematical basis of Einstein's gravitational theory and to the *theory of Hilbert space* as the basis of quantum mechanics[4]. Even Dirac's theory of electrons gave no new impulses to mathematics: the *spinor theory* had already been established by Cartan within the frame work of group representation theory.

4. The advance development of mathematics does not only *facilitate* the development of physics but is *demanded* by the latter. *The increasing distance of the primary concepts of physics from the world of our senses and from the 'physics of daily life',* correctly judged by Planck as an approach to the real world and used as the main argument against positivism, implies an increasing *shift in the methods of research from induction to mathematical construction with subsequent deduction.* Einstein's theory of gravitation can serve as a classical example of this.

5. Logical analysis confirms the conclusions stated above. If a physical theory is given, mathematics is only used to compute the solution of pertinent problems. This *deductive* or *transformative* role of mathematics is entirely unproblematic; it merely serves to draw consequences from the given axioms (fundamental equations) of the theory. Hence, to every physical theory a computer may be constructed which answers any question that is meaningful in the sense of the theory. In accordance with this, mathematics functions as the syntax of the physical language if the latter is formalized.[5] Thus, the 'formalistic' conception of mathematics, founded by Hilbert, is vindicated.

In addition to its deductive role mathematics plays a *constitutive* or *formative role* in physics. It consists in the fact that the *choice of a certain mathematical calculus largely predetermines the fundamental equations* of the [physical] theory in question. Thus, the fundamental equations of *relativistic mechanics*, established by Einstein in a roundabout way, viz., by a *detour* over electrodynamics, do in fact follow from a correspondence axiom once the Minkowski geometry is accepted. Similar remarks apply to the *Maxwell equations* which cannot be deduced from the so-called elementary laws as pointed out by Planck and mentioned in the 'Theses', but also to *Schroedinger's equation* and to *Einstein's gravitational equations*.

At the same time we had to learn that *every essential progress in physical theory is connected with the employment of a new mathematical calculus*, and not merely with new equations in the frame of the old calculus.

Thus, modern physics is bound to demand of mathematics that it should offer suitable calculi the usability of which is to be judged neither *a priori* nor by inductive arguments but by its consequences only. In this point, too, the development of mathematics and the 'formalistic' conception that goes with it fully answer the requirements of physics.

6. As was to be expected, the emancipation of mathematics was not without certain inherent dangers that have become apparent in the meantime. Two of them may be mentioned: (a) the exaggerated trend towards generality of theorems obtained by including all sorts of 'pathological' cases, called *Tendenz zum Purismus* in the 'Theses', and (b) the danger of *overproduction of calculi* which will never be applied.

As far as the *first trend* is concerned there may exist good internal reasons for it; I leave this to the professional mathematicians. As for *overproduction of calculi*, I think this is the price we have to pay for a sufficiently large offer. However, the overproduction of unusable calculi can perhaps be considerably reduced, as I would expect, if the present alienation is overcome.

7. As long as the emancipation process went on, mathematics was bound to alienate itself more and more from physics. Mathematicians had to put their own house in order and to do some completions. Only men of exceptional genius such as Hilbert, Weyl, von Neumann could find time and understanding for physics. [In response to this alienation] physics developed a sort of self-help in the form of what may be called – in contradistinction to theoretical physics – *mathematical physics* and *physical mathematics*, respectively.[6]

Today, the process of emancipation is finished and hence there is no objective need for the alienation to continue. This is confirmed by the recent International Congress of Mathematics (Moscow, 1966); there and on other occasions a certain trend of approach to physics is quite noticeable. The re-approach of mathematics to physics will proceed best if both sides recognize the metascientific facts I have tried to sketch. The decisive fact is the double role of mathematics in physics outlined under 5., which is in full agreement with the 'formalistic' conception of mathematics (Hilbert, Einstein).

8. The correct understanding of mathematics and physics and their relation to one another is impeded or even prevented by a number of popular mistakes. Some of these mistakes are considered in the following.

9. Often mathematics is called *the language of physics.*[7] This is misleading in two respects. First, certain branches of măthematics are also used in non-physical sciences such as economics or military strategy, and some are not used at all. Hence, the above statement would have to be corrected to read: the language of physics is part of mathematics. This statement is still wrong. Many fundamental physical concepts such as 'inertial system' are not at all represented within the mathematical formalism but have to be defined in the metalanguage of the theory. This applies in particular to all concepts in physics that are based on the symmetry properties of the fundamental equations. The correct statement would be that the *syntax* of the [formalized] language of physics is determined completely (and in an essentially unique way) by [the *mathematical formalism*, i.e.] the *mathematical calculus* [*mathematical substructure*] used and the *fundamental equations* [*mathematical superstructure*] of the theory.

10. Some philosophers and mathematicians, refering to the [empirical] origin of mathematics, advocate a *naturalistic conception of mathematics* that would correspond to the relation between Euclid's geometry and empirical geometry (*'abstraction theory'*). Yet the historical origin of a theory, which is largely accidental, says nothing about its true content. Otherwise we would still be obliged to believe in Schroedinger's classical wave theoretical interpretation of quantum mechanics.

11. Sometimes misconceptions concerning the relation between mathematics and physics diffuse from mathematics into physics. An interesting example is found in the 'Theses' (theses nos. 2–4). The matter [which is somewhat involved] is [briefly] as follows.

As is well-known, the emancipation and self-determination of mathematics together with the production of more and more novel calculi led to the demand for a systematic reconstruction of mathematics, viz., its reduction to one or a few basic disciplines. Rival programs originated: 'arithmetization' (Kronecker), reduction to logic (Frege), and – based on the success of Cantor's set theory in mastering the actual infinite (transfinite cardinal numbers) – reduction to set theory. These programs were taken up and developped by *intuitionism* (Brouwer, Heyting), by *logicism* (Russell-Whitehead), and by *mathematical logic*[8]. From the 'formalistic' point of view (Hilbert-Einstein) all these attempts at re-duction are of interest only in so far as they, when successful, permit a reduction of *consistency* [*proofs*]. For the application of mathematics to physics or other branches of science it is the consistency of calculi that matters. Most of the representatives of the named schools, however, consider successful reductions as a kind of 'foundation' of mathematics or as proof of the truth and unavoidability of the 'material'[9] interpreta-tion from which they started.

If now such a 'material' interpretation is combined with a naturalistic conception of the relation between mathematics and physics, this 'material' interpretation is transferred to the foundations of physics, which gives rise to unsolvable paradoxes. If, e.g., one adheres to the set theoretical 'foundation' of mathematics it must appear paradoxical that physics [in the form of quantum mechanics] has succeeded in establishing a mathematical theory of transition probabilities and, in connection with it, of virtual particles and states without inventing an entirely 'novel' (viz. irreducible to set theory) kind of mathematics. This example, taken from the 'Theses', is of quite particular beauty and illuminating power: the quantum mechanical calculus of [transition] probabilities rests indeed on a *partial Boolean* algebra so that a set theoretical interpretation is impossible.[10]

To be sure, this paradox does not prove that a set theoretical founda-tion of Hilbert space is impossible, yet it does show that the 'material' foundation of a calculus is without any significance for physics and, moreover, that in connection with a naturalistic conception of the mathematics-physics relation it leads to unsolvable paradoxes. This is not astonishing if one considers that reduction within mathematics is quite arbitrary: instead of reducing irrational numbers to rational ones (Dedekind) one could equally well define the rational numbers [within an axiomatic theory of real numbers] as real numbers with special properties; this would be even more natural, both from a logical and from the physical point of view. As for set theory in particular, if taken as a calculus it is one amongst a thousand others. Its specific suitability for mathematical logic and metamathematics simply results from the fact that it is the extensional version of predicate logic.

SUMMARY

(a) The emancipation and self-determination of mathematics is a necessary result of developmental laws and irreversible. By it, that advance of mathematics became possible without which modern physics would be impossible.

(b) The alienation of mathematics from physics is a historically limited by-product of this process which can now be overcome.

(c) The essential condition for overcoming it is the recognition by both sides of the actual role of mathematics in physics, in agreement with the 'formalistic' conception of mathematics (Hilbert-Einstein).

(d) Other conceptions of mathematics are arbitrary and not acceptable to modern physics.

(e) Physicists may rightly demand that mathematicians do their part in overcoming the alienation from physics.

NOTES

* Translated from 'Grundsätzliches zum Verhältnis von Mathematik und Physik und zu seiner geschichtlichen Entwicklung', *Monatsberichte der Deutschen Akademie der Wissenschaften zu Berlin* 2 (1967), 356–363.
[1] *Monatsber. Dt. Akad. Wiss. Berlin* **8** (1966), 93–97; in the following quoted as 'Theses'.
[2] A. Einstein, *Ĝeometrie und Erfahrung*, p. 4, Berlin 1941.
[3] In the 'Theses' (thesis 2) this famous dictum is misrepresented by a 'free quotation': instead of *secure (sicher)* we find the term *exact* (That no physical theory is exact since it is but an approximation to reality goes without saying!) In the subsequent commentary the predicate 'exact' is taken to refer to *concepts* although Einstein speaks of *statements* or *theorems* (Saetze) of mathematics. These two 'corrections' of Einstein's dictum do not cancel one another: Einstein never advocated the absurd view that [as the 'Theses' have it] "physical concepts cannot be coordinated to the axiomatically sharply defined mathematical concepts since the former must necessarily possess a certain latitude of definition". (Even Bohr, whose view on this matter has possibly been mistaken for Einstein's view, merely speaks of a certain complementarity between sharpness of definition on the one hand and intuitive meaning and common usage on the other hand.)

Incidentally, it would be a mistake to believe mathematical concepts were sharply (uniquely) defined: this would not even be true for a 'complete' axiomatic system which still admits of isomorphic transformations, not to speak of ordinary axiomatic systems that admit of non-isomorphic realizations or models. Mathematical axioms determine the syntactic properties of a concept, no more and no less. *'Sharp', in a mathematical calculus, are not the concepts but the rules of formation and transformation.*
[4] In the case of quantum mechanics, certain fundamental equations were found without knowledge of Hilbert space, yet their physical interpretation remained unclear and controversial until the mathematical foundations were clarified (von Neumann, Born).

[5] M. Strauss, 'Mathematics as Logical Syntax – A Method to Formalize the Language of a Physical Theory', *Erkenntnis* **7** (1938), 147–153; [Chapter VI of this volume].

[6] Attempt at definition: A *theoretical physicist* does physics *although* he needs mathematics to do so. A *mathematical physicist* does physics *because* he needs mathematics to do so. A *physical mathematician* does mathematics *although* his mathematical work is useful in physics.

[7] Of course, Galileo was *historically* and *intentionally* right in using this phrase: it emphasized the *indispensability* (then in doubt) of mathematics for physics.

[8] *Mathematical logic*, in the strict sense, is logic treated mathematically, or – what amounts to the same – the mathematical theory of calculi that admit of logical interpretation. Today, the term is often used in a wider sense, meaning logic of mathematics or metamathematics.

[9] 'Material' not in the sense of physical or social science but in the sense of non-formal interpretation.

[10] M. Strauss, 'Zur Begruendung der Statistischen Transformationstheorie der Quantenphysik', *Sitz.-Ber. Berl. Akad. Wiss., phys.-math. Kl.* **27** (1936), 382–398; [Chapter XVI of this volume], F. Kamber, 'Die Struktur des Aussagenkalküls in einer physikalischen Theorie', *Nachr. Akad. Wiss. Goettingen* **10** (1964), 103–124.

CHAPTER VI

MATHEMATICS AS LOGICAL SYNTAX – A METHOD TO FORMALIZE THE LANGUAGE OF A PHYSICAL THEORY*

I. INTRODUCTION

I intend to explain a method of formalizing the language of a given physical theory. The essential part of this method consists in introducing *descriptive predicates uniquely related to mathematical entities* and hence *using the mathematical formalism* of the given theory *to establish the logical syntax of* the formalized language.

Instead of describing the method in abstract terms I shall illustrate it by *two examples: classical* and *quantum mechanics*. It may be said in advance that by formalizing the language of quantum mechanics we obtain a *syntactic definition of quantum mechanical complementarity* already suggested in previous papers[1].

II. CLASSICAL MECHANICS

Using the Hamiltonian form of classical mechanics, every subset M of the total phase space E can be interpreted as follows: the phase point of a certain mechanical system s lies at a certain moment t in M. This suggests the introduction of sentential functions

$$p_M(s, t) \tag{1}$$

having the indicated meaning. In this way a one-one correspondence is established between the atomic predicates 'p_M' and the subsets of E.

We can now define *predicate connectives* as follows:

$$(C\ I) \quad \left| \begin{array}{c} \sim p_M =_{df} p_{\bar{M}} \\ p_M \wedge p_{M'} =_{df} p_{M \cap M'} \\ p_M \vee p_{M'} =_{df} p_{M \cup M'} \\ p_M \supset p_{M'} =_{df} p_{\bar{M} \cup M'} \end{array} \right| \tag{2}$$

By these definitions the Boolean algebra of the set calculus is transferred to the predicate calculus. At the same time these definitions ensure that – in contradistinction to other languages – every compound predicate is equivalent to (replaceable by) an atomic predicate.

Hence we can formulate *transformative rules* as follows:

(C TR I) Any sentence equivalent to $p_M(s, t)$ is *valid* iff $M = E$.

(C TR II) Any sentence equivalent to $p_{M_1}(s, t)$ is a *consequence* of any sentence equivalent to $p_{M_0}(s, t)$ if $\bar{M}_0 \cup M_1 = E$ (or, equivalently, if $M_0 \subseteq M_1$).

The physical content of the equations of motion

$$\dot{q}_i = (\partial H / \partial p_i) \qquad \dot{p}_i = - (\partial H / \partial q_i) \tag{3}$$

can be formulated as an *additional transformative rule*. To this end we use the fact that the equations of motion define a mapping of the phase space onto itself. Let $V(t)$ be the operator of this mapping and

$$M^\tau =_{df} V(\tau) M. \tag{4}$$

The equations of motion are then equivalent to the rule

(C TR III) $p_M(s, t) \rightleftharpoons p_{M^\tau}(s, t + \tau).$ \hfill (5)

Alternatively, we may introduce new predicates by

$$p_M^\tau(s, t) =_{df} p_M(s, t + \tau) \tag{6}$$

and use the rule

(C TR III′) $p_M \rightleftharpoons p_{M^\tau}^\tau.$ \hfill (7)

III. QUANTUM MECHANICS

In quantum mechanics the classical phase space E is replaced by a Hilbert space $\mathfrak{H}$. Since the latter is *linear vector space* a physical interpretation cannot be given to *any* subset of $\mathfrak{H}$ but only to *closed linear subsets* (*subspaces*) $\mathfrak{M}$ of $\mathfrak{H}$.

Thus it may appear that in order to formalize the language of quantum mechanics by the method used above we simply have to use *subspaces* instead of *subsets*. This is indeed the idea underlying Birkhoff and v. Neumann's 'logic of quantum mechanics'. However, this idea is not correct because there is an *ambiguity* in the choice of the mathematical entities that are to be correlated to physical predicates. This ambiguity results from the fact that there is a *one-one correspondence* between *subspaces* and *projectors* in $\mathfrak{H}$, to every $\mathfrak{M}$ belonging exactly one projector $\Pi_{\mathfrak{M}}$ and vice versa. This ambiguity would have no consequences for our problem if the two calculi – that of the subspaces and that of the projectors – were isomorphic. In fact, however, the calculus of the subspaces is an instance of *non-distributive* orthocomplemented *lattice algebra* while the calculus of projectors is an instance of *partial Boolean algebra*, i.e., roughly speaking, an algebra consisting of Boolean sub-algebras that do not add up to a Boolean algebra. They don't add up to a Boolean algebra because the *connectives* are *restricted* to elements of the *same* subalgebra. Hence a *logic* with *partial* Boolean algebra may also be called a *logic with restricted sentential* (or predicate) *connectives*. Predicates that cannot be connected to form a compound predicate are *complementary* in the sense of N. Bohr. Thus, a predicate logic with partial Boolean algebra can be considered as a *logical codification* of Bohr's ideas on complementarity.

[From the semantic point of view, a logic with partial Boolean algebra can be considered as *two*-valued while a logic with *non*-distributive lattice algebra implies at least *three* truth values. The latter follows from the fact that *two*-valuedness *implies* the distributive law – a fact apparently not noticed by the adherents of the BIRKHOFF-V. NEUMANN [2] logic.]

Thus, there are good reasons for choosing *projectors* rather than *subspaces* as the mathematical entities to be correlated to physical predicates.

[For the sake of generality we shall not place any restriction on the class of projectors admitted. However, in most cases of physical interest the projector is related to a so-called 'observable', or rather to a Hermitean operator A representing an 'observable'. (An 'observable' is, logically speaking, a dispositional property, more precisely: a stochastic mode of reaction.) Indeed, if $E^A(a)$ is the resolution of identity (unity) belonging to A, i.e., if

$$A = \int a \, dE^A(a), \tag{8}$$

the projectors of physical interest are of the form

$$\prod_{(a_1, a_2)}^{A} =_{\mathrm{df}} E^A(a_2) - E^A(a_1) \tag{9}$$

These projectors represent the dispositional property that the system concerned when acted upon by a stochastic state changer of class A takes, with a certain probability, a state where the value of A lies in the interval (a_1, a_2). [In the original text, only these projectors have been considered.]

The formalization of the quantum mechanical language now proceeds as follows.

We introduce primitive sentential functions of the form

$$p_{\Pi}(s, t) \tag{10}$$

having the following meaning: the state vector of system s lies at the moment t in the subspace $\mathfrak{M}_{\Pi}$. In case the projector Π is of the form (9) this means: the system s is at t in a state where the value of A lies in the interval (a_1, a_2).

Next we define *predicate connectives* as follows:

$$
\begin{aligned}
\dot{p}_{\Pi} &=_{df} p_{\bar{\Pi}} \\
p_{\Pi} \stackrel{.}{\wedge} p_{\Pi'} &=_{df} p_{\Pi \cdot \Pi'} \quad && \text{if } \Pi \cdot \Pi' = \Pi' \cdot \Pi \\
p_{\Pi} \stackrel{.}{\vee} p_{\Pi'} &=_{df} p_{\Pi \dot{+} \Pi'} \quad && \text{(otherwise undefined)} \\
p_{\Pi} \stackrel{.}{\supset} p_{\Pi'} &=_{df} p_{\bar{\Pi} \dot{+} \Pi'}
\end{aligned}
\tag{11}
$$

On the right-hand side we have used the definitions

$$\Pi \dot{+} \Pi' =_{df} \Pi + \Pi' - \Pi \cdot \Pi' \tag{12}$$

$$\bar{\Pi} =_{df} I - \Pi \qquad (I =_{df} \Pi_{\mathfrak{H}}) \tag{13}$$

By the definitions (11) the projector algebra $[I, \{\Pi\}, \dot{+}, \cdot, -]$ is represented isomorphically by the predicate algebra $[p_I, \{p_{\Pi}\}, \dot{\vee}, \dot{\wedge}, \dot{-}]$. The condition $\Pi \cdot \Pi' = \Pi' \cdot \Pi$ is the necessary and sufficient condition for the compound operators $\Pi \cdot \Pi'$ and $\Pi \dot{+} \Pi'$ to be projectors, and hence for the compound predicate expressions to be predicates. If this condition is not satisfied the two predicates p_{Π}, $p_{\Pi'}$ are *inconnectible*.

Note that by virtue of the definitions (11) every compound predicate is equivalent to (replaceable by) an atomic predicate.

Hence we can formulate *transformative rules* as follows:

(Q TR I)   Any sentence equivalent to $p_{\Pi}(s, t)$ is *valid* iff $\Pi = I$.

(Q TR II)   Any sentence equivalent to $p_{\Pi_1}(s, t)$ is a *consequence* of any sentence equivalent to $p_{\Pi_0}(s, t)$ if $\bar{\Pi}_0 \dot{\cup} \Pi_1 = I$.

The physical content of the Schroedinger equation

$$(H + ih(d/dt))f = 0 \tag{14}$$

can again be formulated as an additional transformative rule. To this end we use the fact that the Schroedinger equation defines a mapping of $\mathfrak{H}$ on itself, the operator of this mapping being

$$U(\tau) = \exp(i\tau H/h). \tag{15}$$

Let

$$\mathfrak{M}^\tau =_{df} U(\tau)\,\mathfrak{M} \tag{16}$$

and hence

$$\Pi_{\mathfrak{M}\tau} = U(\tau)\,\Pi_{\mathfrak{M}}\,U(\tau)^{-1}, \tag{17}$$

or, equivalently,

$$\Pi^\tau = U(\tau)\,\Pi\,U(\tau)^{-1}. \tag{18}$$

The equations of motion are then equivalent to the rule

(Q TR III) $\quad p_\Pi(s, t) \rightleftharpoons p_{\Pi^\tau}(s, t + \tau)$ \hfill (19)

Alternatively, we may introduce new predicates by

$$p_\Pi^\tau(s, t) =_{df} p_\Pi(s, t + \tau) \tag{20}$$

and use the rule

(Q TR III') $\quad p_\Pi \rightleftharpoons p_{\Pi^\tau}^\tau.$ \hfill (21)

Finally, we wish to formulate the stochastic law of quantum mechanics as a transformative rule for our Q-language.

To this end we introduce a *probability functor*

$$\mathrm{prob}_2(p_\Pi; p_{\Pi'}) \tag{22}$$

which is supposed to be defined for any ordered pair of predicates $(p_\Pi; p_{\Pi'})$, whether connectible or not. The syntax of 'prob$_2$' can then be fixed by adopting REICHENBACH's axioms[3] in the 'W-formulation' *for all Boolean subalgebras* of our basic algebra $[p_1, \{p_\Pi\}, \wedge, \vee, \dot{-}]$.

The stochastic law of quantum mechanics can now be formulated as follows:

(Q TR IV) All sentences equivalent to
$$\mathrm{prob}_2(p_{\Pi_0}; p_{\Pi_1}) = r_{01}$$

are valid iff

$$r_{01} = Tr\ \Pi_0 \Pi_1 / Tr\ \Pi_0 \,,$$

'Tr' being the trace operator.

The time-dependence of these probabilities can be accounted for by the simple rule

(Q TR V)   In Q TR IV, 'Π_1' may be replaced everywhere by 'Π_i^t'.

The semantics of 'prob$_2$' will be dealt with elsewhere.

IV. [CONCLUDING REMARK]

[It will be noted that both the (implied) rules of sentence formation and the given transformative rules employ only *mathematical* criteria. This establishes our claim that the mathematical formalism of a physical theory may be considered as the logical syntax of the formalized theoretical language.]

NOTES

* Revised reprint from *Journal of Unified Science*, (*Erkenntnis*) **7** (1938) 147–53.
[1] M. Strauss, 'Zur Begründung der statistischen Transformationstheorie der Quanten-physik', *Sitzungsberichte der Berliner Akademie der Wissenschaften, phys.-math. Kl.* **27** (1936) 382–98; 'Komplementarität und Kausalität im Lichte der Logischen Syntax', *Erkenntnis* **6** (1937) 335–9; 'Mathematische und logische Beiträge zur quantenmechani-schen Komplementaritätstheorie', Dissertation Prague 1938 (unpublished).
[2] G. Birkhoff and J. v. Neumann, 'The Logic of Quantum Mechanics', *Annals of Mathematics* **37** (1936) 823–43.
[3] H. Reichenbach, *Wahrscheinlichkeitslehre*, Leiden 1935, p. 118.

CHAPTER VII

IS THE FREQUENCY LIMIT INTERPRETATION OF PROBABILITY A MEANINGFUL IDEALIZATION?*

It is a well-known fact that idealizations play a large role in science. Thus, in celestial mechanics the planets are treated as mass points, and in the physics of ordinary matter we use the methods of the differential and integral calculus although we know very well that ordinary matter is not continuous and that integrals taken over a volume of matter should be replaced by sums taken over molecules.

This fact has been used by various authors[1] as an argument in favour of the frequency limit interpretation of probability. This interpretation, so the argument goes, is an idealization of the same kind as other idealizations used in science and especially in physics.

In my opinion, this argument is either *wrong* or *unproved*, depending on its precise interpretation. Furthermore, the argument blurs the difference between *wrong* and *meaningless* statements. [[...]]

To start with, it must be pointed out that all idealizations used in physics are *approximations*. Thus, in the case of celestial mechanics the numerical results would hardly be affected if the finite extension of the planets, deviation from spherical shape, and possible inhomogeneities in mass distribution were taken into account. In the majority of celestial problems the numerical differences would be too small for observational detection, and in the minority of problems where this is not so, e.g. in the Earth-Moon problem (tides!), we can replace the idealization by a weaker one, i.e., we can use a better approximation. Similar remarks apply to the use of the differential and integral calculus in the physics of ordinary matter.[2] [[ ]]

If we now ask the question 'What is approximated by the frequency limit?', we are clearly at a loss to answer. If in any probability statement the term 'probability' is replaced by 'limit of relative frequencies' the statement becomes untestable even when 'testable' means 'approximately testable', since a finite sequence, no matter how long, says nothings about existence and value of a limit. [[...]] Thus, if the term 'idealization' in the argument under discussion refers to the comparison of *statements* the argument is certainly wrong.

If, in spite of this, the argument is being presented again and again by respectable scientists we may suspect that something else is meant.

This 'something else', I submit, is the following statement. The *calculus* of probability based on a definition of 'probability' as limit of relative frequencies is an idealizing approximation to a *calculus* of probability not presupposing the existence of such limits. This is to say: if we start with given probability statements and deduce from them other probability statements according to the rules of one or the other calculus, the deduced probabilities will be approximately equal for the two calculi.

If a statement of this kind turns out to be true for a certain limit-free probability theory we could use the (presumably more simple) limit theory for deducing new probability statements from given ones and still maintain a limit-free interpretation in the sense of the limit-free theory.

Quite obviously, in this second interpretation the argument under discussion is unproved and could be proved only by developing a limit-free theory of probability. Thus, the proponents of the limit theory of probability should be the first persons interested in developing a limit-free theory of probability.[3]

NOTES

* Translated from 'Ist die limes-Theorie der Wahrscheinlichkeit eine sinnvolle Idealisierung?', *Synthese* **5** (1946/47) 90–1. (The paper was written in 1939 in The Hague at the suggestion of Otto Neurath.)

[1] E. Nagel, *Principles of the Theory of Probability, International Encyclopedia of Unified Science*, Vol. I, No. 6, Chicago 1939 (in particular p. 54). – R. v. Mises, *Wahrscheinlichkeit, Statistik und Wahrheit*, Berlin 1928. – C. C. Hempel, *Unity of Science Forum*, September 1938.

[2] As to the approximating character of science, cf. V. F. Lenzen, *Procedures of Empirical Science, International Encyclopedia of Unified Science*, Vol. I, No. 5, Chicago 1938.

[3] Outlines of such a theory are suggested in M. Strauss, 'Problems of Probability Theory in the Light of Quantum Mechanics, Part III', *Unity of Science Forum*, April 1939, 65–71.

CHAPTER VIII

PROBLEMS OF PROBABILITY THEORY
IN THE LIGHT OF QUANTUM MECHANICS*

PART I

0. *Introduction I*

Through the work of modern [mathematical] physicists and logicians[1] the clarification of the probability concept as used in empirical science has reached a high degree of perfection. Some questions, however, concerning the *form* of probability statements as well as their *meaning* are still under discussion[2]. Therefore it may be of some interest to investigate what contribution to the solution of these problems can be obtained by the logical analysis of a theory the fundamental laws of which are probability laws.

At first sight, quantum mechanics may appear to be unable to make any further contributions beyond those already given by classical statistical mechanics. Yet in this context a very important *difference* between quantum mechanics and any other probabilistic theory known so far must be emphasized. Apart from quantum mechanics, any probabilistic theory starts from certain fundamental probability statements (probabilistic axioms) and proceeds to other probability statements (mostly more suitable for observational testing) by *explicit* use of the probability calculus. It is just this fact which allows to separate the well-established *transformative rules* from the more difficult questions of *form* and *meaning*. Quantum mechanics, on the other hand, is given to us as a certain *mathematical formalism supplemented by a probabilistic interpretation*; it therefore contains a probability theory *implicitly*. Hence we can hope to get information on the questions mentioned by analysing this implicit theory.

It should be noted, however, that this information cannot be 'picked up' directly from the mathematical formalism of quantum mechanics. 'Probability' is a *descriptive* term, not a logical or mathematical one; hence it cannot occur in the mathematical formalism. In order to get the desired information we must *formalize* the quantum mechanical language beforehand. Though formalization may be carried out in different ways, the essential features are always determined by the mathematical formalism of the physical theory concerned. Thus, the desired information *does* follow from the mathematical formalism, though indirectly.

The formalization presupposed in the following is the one carried out in a previous paper [3]. [It will suffice here to say that

(i) the *atomic predicates* of the formalized language form a *partial* Boolean algebra, i.e., roughly speaking, an algebra consisting of Boolean subalgebras, the elements of *different* such subalgebras being *inconnectible* by the predicate connectives,

(ii) every *compound* predicate is equivalent to (replaceable by) an *atomic* predicate, and

(iii) the *probability functor* 'prob' is a *two*-place functor over the cross-product of any two of the Boolean subalgebras mentioned.]

1. *Form of Probability Statements*

In this section we discuss the *logical form* of probability statements.

1. Let us begin with the question whether the word 'probability' shall be used as a word of the physical *object-language* or as a word of its *meta-language*; obviously, this question must be decided before other questions concerning the form of probability statements can be discussed.

As for this question, quantum mechanics does not give us an answer; we can formalize quantum mechanical language in the one and also in the other way. This fact agrees with the modern view emphasized generally by CARNAP [4] and, as for the point in question, by REICHENBACH [5].

Of course, if we banish the word 'probability' from the physical object-language we *lose* the possibility of expressing the probabilistic laws of physics in that language and *gain nothing* because the problems of *form* and *meaning* would merely be transferred to the next semantic level. Hence we take the work 'probability' as belonging to the physical object-language.

2. The next question to be considered is this: is 'probability' a *property* of an event (or class of events) or a *relation* between events or classes of events? We will come back to this point later. In the first case we call the probability *absolute*, in the second case *relative*.

The modern view on this question established by v. MISES [6] and REICHENBACH [7] is that all probabilities are *relative*.

Quantum mechanics confirms this view. We need not refer to formalization to show this. Quantum mechanics is well known to give a probability for the result of a measurement if and only if the result of a precedent measurement [or the 'preparation' of the physical system concerned] is known.

3. Although the relational character of 'probability' is now widely admitted, the question 'relation between what kind of objects' is still under discussion. [This is not surprising considering that different kinds of objects such as *events* and *properties* can be represented formally by the *same algebra* so that from the mathematical point of view there is no difference between them. Incidentally, this example shows that we cannot expect any information on the questions here discussed from purely mathematical theories of probability.]

Now in the formalized language of quantum mechanics the arguments of the two-place 'prob' functor are *predicates of equal type*; hence the probabilistic relation in quantum mechanics is one between *properties of equal type*, i.e., properties ascribable to the *same* object, and of course to any similar object. [In quantum mechanics, these properties are *dispositional properties* (modes of *stochastic* reaction, to be precise) while in classical statistical mechanics and most other probabilistic theories they are ordinary properties. This is reflected formally by the difference between partial Boolean and Boolean algebra of the predicate calculus, but this makes no difference to the question here discussed.]

Thus, we take as the correct form of a probabilistic statement not the form

(I) $\quad \text{prob}(P_0(s_0, t_0); P_1(s_0, t_1)) = p$

but

(II) $\quad \text{prob}(P_0; P_1) = p$.

'$\text{prob}(P_0; P_1)$' may be read: 'the probability that any object having property P_0 also has the property P_1' or 'the probability of property P_1 to coexist with property P_0'.

Usually, the properties P_i are defined by sentential functions of the form shown in (I), with 's' for 'physical system' and 't' for 'time'. Now it is quite clear that a probabilistic law applies to *all* physical systems of the same kind so that the individual constant 's_0' cannot occur in it.

Furthermore, all physical laws are supposed to be independent of time so that the individual constants 't_0' and 't_1' cannot occur in them either. Time *intervals* τ *can* of course occur in physical laws, but this can be accounted for by admitting for 'P_1' in (II) predicates defined according to the scheme

$$P^\tau(s, t) =_{df} P(s, t + \tau).$$

4. We now compare form (II) with other forms that have been suggested.

Form (II) occurs in REICHENBACH's book [7], but there it is introduced as an *abbreviation* for a rather complicated expression which can be written in the form [8]

(III) $\mathrm{prob}(S_0; P_0, P_1) = p_0$

where 'S_0' means a *sequence* of events. Using the statistical interpretation, (III) could be read: 'The relative frequency of those elements of S_0 which have the properties P_0 and P_1 among the elements having property P_0 is equal to p_0'.

Form (III) has been objected to by HEMPEL [9] on the ground that no physical theory contains any reference to a sequence S_0. This objection is quite correct; it is also confirmed by quantum mechanics. In spite of this, HEMPEL [9] believes that introduction of a sequence term into probability sentences is necessary because without it, in his opinion, the mathematical theory of probability could not be applied.

As the solution of this 'dilemma' HEMPEL [9] proposes the form

(IV) $(S) [C(S) \supset \mathrm{prob}(S; P_0, P_1) = p_0]$

where 'C' means 'a group of conditions characterising those sequences in which P_1 has the probability p_0 with respect to P_0'.

This proposal does not solve the 'dilemma': we can again object to the form (IV) that no physical theory contains any reference to *such* a condition. Besides, if the condition C has the meaning ascribed to it the sentence (IV) would be *analytically true* whereas a physical probability statement is synthetic and hence logically indeterminate.

The solution of Hempel's 'dilemma' consists in the fact that there is no such dilemma at all. Every probability statement [occurring in science] can be written in the form (II), and all we need in a general theory of probability follows from *axioms concerning this form*, as given by REICHENBACH[10] in his 'W-formulation'. These axioms are tautologically ['identically'] fulfilled by any frequency interpretation, whether it be finite or infinite. But from this it does not follow that frequency interpretation can be introduced into the formalized language of a physical theory. Indeed, a frequency interpretation must refer to *real* [and *finite*] ensembles while a probabilistic law refers to a *virtual* [and infinite] ensemble [as is well-known from Gibbs' Statistical Mechanics]. It is just the insufficient distinction between *syntax* and *meaning* [and between *meaning* and *testing*] which leads to dilemmas and paradoxes.

[To be sure, *equality in form* of all probability statements is a *necessary* condition for the existence of a *general* (comprehensive) theory of probability. To obtain a *necessary and sufficient* condition we must add the *further condition* that the *predicate algebras*, too, are the *same* for all probabilistic theories. This further condition is *not* satisfied, as the case of quantum mechanics shows. Hence we need (at least) *two* somewhat different theories of probability, axiomatizing two different concepts of probability, $prob_1$ and $prob_2$, corresponding to Boolean and partial Boolean algebra, respectively. If '$prob_1$' means a probability of coexistence of properties, '$prob_2$' means a probability of induced stochastic transitions between states. However, in spite of this fundamental difference in the semantics of '$prob_1$' and '$prob_2$', the *formal* theory of $prob_2$ will contain the *formal* theory of $prob_1$ as a *degenerate case* – in the same sense, and for the very reason, that Boolean algebra is a degenerate case of partial Boolean algebra.]

2. *Introduction II*

In Part I we dealt with the *form* of probability statements. In this and the following Part the rules of *transformation* and the *meaning* of probability statements will be discussed, both from the viewpoint of quantum mechanics.

In the older investigations on probability the *opposite* way was taken: the starting point was a *material definition* of 'probability' [or what was supposed to be such a definition], and the *transformative rules* were *de-*

rived as consequences of the given definition. This has been the procedure in the classical papers on probability and it still *is* the procedure in v. Mises' theory.

At first sight, this procedure looks quite natural. Unfortunately, a satisfactory material definition of 'probability' has never been given. The classical definition based on the concept of 'equal possibilities' is circular [or nearly so] since 'equal possibilities' is only another word for 'equal probabilities'. As for v. Mises' definition, not even its consistency can be taken for granted. Modified definitions proposed by other authors have been objected to for other reasons.

Thus it seems to be more adequate to the present situation to begin with formal rules of formation and transformation for an undefined probability concept, allowing to derive the customary theorems needed in applications, and to discuss the possibilities of material interpretation afterwards. This way, followed by Reichenbach with great success, has the further advantage that the formal theory can be used as a basis for discussing the *meaning* problem the solution of which is of course partly determined by the formal rules of formation and transformation.

[Apart from these methodological reasons there is a more fundamental reason for adopting the procedure outlined above: 'probability' is an *irreducible* theoretical concept so that the axiomatic method of definition is the only one applicable.]

As the *sentential form* of a probability satement is presupposed in both the *transformative* and the *semantic* rules we start with some supplementary remarks on the *form* of probability statements.

3. *Form of Probability Statements – Continued*

As shown in Part I, any quantum mechanical probability statement can be written in the form

(II) $\text{prob}(P_1 ; P_2) = p_0$.

Furthermore, we have found no reason for supposing that a probability statement of any other branch of science would require a different form. Hence we shall presuppose form (II) in the following.

There remain however some other questions concerning the form of probability statements which we shall discuss now.

1. Let us begin with the question whether or not the two *predicates*

in the standard form (II) must be of *equal type*. In the discussion of this question we presuppose the predicates to be one-place predicates, i.e., defined by one-place sentential functions. This is no real restriction since an n-place sentential function can be considered as a one-place function for an ordered n-tuple of variables. Let 'P$_1$' be defined by the sentential function 'P$_1$(x)' with x ε M$_1$, and 'P$_2$' by 'P$_2$(y)' with y ε M$_2$. Then 'P$_1$' and 'P$_2$' are said to be of *equal type* iff M$_1$ = M$_2$. [If 'P$_1$' is a n-place predicate, M$_1$ can be written as n-fold cross-product.]

[Now the first thing to notice is that *predicates of equal type* are *implied* by any system of axioms or transformative rules equivalent to those given by Reichenbach: the predicate connectives occurring there are defined (and indeed definable) for predicates of equal type only, the definition scheme being (P$_1$ c P$_2$) (x) =$_{df}$ P$_1$(x) c P$_2$(x), where 'c' stands for any of the connectives.]

The second point to notice is the fact that *equality of type* does *not guarantee* that the two predicates denote properties ascribable to the same real object, a counter-example being the predicates P$_1$(x) =$_{df}$ Q$_1$(s$_1$, x) and P$_2$(x) =$_{df}$ Q$_2$(s$_2$, x) where 's$_1$' and 's$_2$' mean different physical systems and 'x' may be the time variable. If we wish to exclude such counter-examples we must demand that the predicates P$_1$ and P$_2$ are not only of *equal type* but *also* of the *same (semantic) kind*.

Thus, a probability expression may be said to have *standard form* if it has form (II) with 'P$_1$' and 'P$_2$' for predicates of equal type, and to have *full standard form* if it has standard form and 'P$_1$' and 'P$_2$' are of the same (semantic) kind.

Now the probability expressions in the formalized quantum mechanical language constructed by the author[11] *have* full standard form, and there is no reason to suppose that the probability expressions of any other scientific theory cannot be brought into this form. This greatly facilitates the discussion of the meaning problem.

2. Next we turn to Reichenbach's 'rule of existence'. This rule claims[12] that a probability relation exists (i.e., that (II) is a sentence), if its numerical value ('p$_0$' in (II)) is determined by the values of other probability relations according to the given arithmetical axioms. For instance, if the expressions

(1)   prob (P$_1$; P$_2$) = p$_1$

(2) $\text{prob}(P_1 \wedge P_2; P_3) = p_2$

are sentences, then – so the rule claims – the expression

(3) $\text{prob}(P_1; P_2 \wedge P_3) = p_3$

is also a sentence because p_3 is determined by p_1 and p_2 according to the general theorem of multiplication ($p_3 = p_1 \cdot p_2$). Thus, the 'rule of existence' is indeed a *general formative* rule, although it refers to axioms (written as functional equations for 'prob'). However, it only *'transfers existence'* and may hence be called a *relative rule of formation.*

In formalized quantum mechanical language this rule of existence has to be *restricted* by the condition that the compound predicate expressions occurring in the 'conclusion' (i.e., the prob expression to which Reichenbach's rule transfers existence) *are* predicates; [this condition is equivalent to the condition that the compound predicate expression is replacable by an atomic predicate]. This restriction is a direct consequence of the fact that the predicate connectives of the formalized quantum mechanical language belong to a *partial* Boolean algebra. The inconnectible predicates represent disposional properties that are *complementary* in the sense of N. Bohr.

3. There remains the question whether *any* expression of form (II) represents a probabilistic statement. The answer to this question will naturally depend on the semantic rules fixing the meaning of 'prob'. Hence we shall not argue this question here any further [except for suggesting that the answer is 'yes' provided the sentential expression is of *full* standard form].

4. *Rules of Transformation for Probability Statements*

A rule of transformation tells us under what formal conditions a sentence is a *consequence* of other sentences. [Axioms, when considered as consequences of the empty sentential class, are a special class of transformation rules]. If a theory is presented in axiomatic form the transformative rules of formal logic are implied without being mentioned.

[In mathematics, an *equation* is usually treated as both a *sentence* of the object-language and a *replacement rule*, i.e., a transformative rule of special kind. In the *classical* theory of probability [prob_1 theory] we could do likewise, and then Reichenbach's 'rule of existence' would be superfluous. For the quantum mechanical language (as formalized by the author) Reichenbach's rule of existence has to be *restricted*, as pointed out above. This implies that equations between prob$_2$-expressions *cannot* in general be treated as replacement rules, e.g., in the general multiplication theorem

(M) $\text{prob}(P; P' \wedge P'') = \text{prob}(P; P') \times \text{prob}(P \wedge P'; P'')$

the compound expression 'P' $\wedge$ P''' may *not* be a predicate so that the left-hand side may be meaningless, even if the right-hand side is meaningful. For this reason Reichenbach's method of *not* taking equations between prob expressions as replacement rules becomes indispensible in prob$_2$ theory.]

Besides the general multiplication theorem (M) Reichenbach takes as axioms the general theorem of addition

(A) $\text{prob}(P; P' \vee P'') = \text{prob}(P; P') + \text{prob}(P; P'') -$
$$- \text{prob}(P; P' \wedge P'')$$

and the normalization axioms

(N 1) $\text{prob}(P; P' \vee \overline{P'}) = 1$

(N 2) $\text{prob}(P; P' \wedge \overline{P'}) = 0$

(N 3) $0 \leqslant \text{prob}(P; P')$.

[Here, the system $[\{P\}, \vee, \wedge, -]$ is of course taken to be a Boolean algebra, hence the axioms can be taken over for any Boolean subalgebra of a partial Boolean algebra. Moreover, with the *restricted* rule of existence it can be taken over for the partial Boolean algebra itself; in this case the connectives '$\vee$', '$\wedge$', '$-$' should be replaced by the corresponding connectives '$\dot\vee$', '$\dot\wedge$', '$\dot-$' of partial Boolean algebra.]

From the mathematical point of view, (M) and (A) are *functional equations* for an unknown two-place function 'prob' defined over $\mathfrak{L} \times \mathfrak{L}$, where '$\mathfrak{L}$' means 'Boolean algebra'. [The general solution of these equations has been given by Kolmogoroff; it reads

$$\text{prob}(P; P') = f(P \wedge P')/f(P) \tag{1}$$

where f is any 'additive function':

$$f(P \vee P') = f(P) + f(P') \quad \text{for} \quad P \wedge P' = 0. \tag{2}$$

We said 'general solution' though it would be more correct to say that an unknown two-place function has been *reduced* to an unknown one-place function. If instead of $\mathfrak{L}$ we have a partial Boolean algebra, $\mathfrak{L}^*$, the 'general solution' would obviously read

$$\text{prob}(P; P') = f(P \overset{.}{\wedge} P')/f(P) \tag{3}$$

with

$$f(P \overset{.}{\vee} P') = f(P) + f(P') \quad \text{for} \quad P \overset{.}{\wedge} P' = 0.] \tag{4}$$

Now in quantum mechanics the (dispositional) predicates P are represented by *projectors* (projection operators) Π in a Hilbert space. Hence prop$(P; P')$ must equal a numerical function of Π, Π'. Indeed, we have here

$$\text{prob}(P; P') = \text{Tr}(\Pi \overset{.}{\wedge} \Pi')/\text{Tr}(\Pi) \tag{5}$$

[and this is indeed of the form (3)]. 'Tr' is the trace operator.

Thus, quantum mechanics does not tell us anything new concerning the *transformative* rules of probability theory. [It raises however the question whether there exist other mathematical representations of partial Boolean algebra, besides projectors in a linear vector space. If this should not be the case, (5) would be *the only solution*, in marked contrast to the case of prob$_1$ theory (Boolean algebra) where the 'measure function' f remains undetermined].

PART III

5. *Introduction III*

In this final part we shall deal with the question what is *meant* by a probabilistic statement occurring in science.

For a treatment of this question we have two independent sources: the *syntax* of the probability functor 'prop' and the *statistical methods* used for *testing* [or even *suggesting*] probabilistic laws.

To be sure neither of these two sources is sufficient for an unambiguous solution of our problem. However, some progress may be expected by considering both sources simultaneously. We can infer from scientific practice that only some kind of *frequency interpretation* can give us the actual [pragmatic] meaning of probabilistic statements, and we can infer from the standard form (II) [assumed to be *full* standard] that probability is a *relation between properties* of the same semantic kind, i.e., properties that can meaningfully be ascribed to the same physical object.

6. *The Meaning of Probabilistic Statements*

1. Before explaining the interpretation here proposed I shall deal with the well-known 'limit interpretation' which seems to have found the most general approval [and which can still be found in textbooks].

The 'limit interpretation' has to be *rejected* for two independent reasons.

First, it *cannot be applied* unless the set of events we are dealing with is *ordered* as a sequence: probability is regarded on this interpretation as the limit of relative frequencies in an *infinite sequence*. Even if we concede [as we must] that all sets [or 'ensembles'] to which probabilistic laws refer are *infinite*, we are not allowed to suppose them to be *ordered*. Of course, each infinite set [with cardinal number aleph zero] can be ordered, but the order so obtained, and hence the limiting value of the relative frequency, would be *man-made* (not *natural*). [Moreover, the predicate 'P_0' in the standard form (II) defines a *virtual ensemble*, namely the *class* of *all* objects having property P_0. The cardinal number of such a class, if it can be defined and determined at all, will in general be a *Tarski unattainable* cardinal number or at least that of the continuum, so that there is not even the logical possibility of sequential order.] Thus, in general the 'limit interpretation' is *inapplicable*.

Second, even in the exceptional cases where the 'set of events' is given by nature in the form of a *sequence* so that the limit interpretation is logically applicable, this interpretation would render the probabilistic statements *neither confirmable nor disconfirmable*: no finite set of observations can decide in favour or disfavour of a limit statement. (Only if we presuppose the sequence to have a certain structure – or lack of structure – e.g. to be 'normal' (irregular) does the limit interpretation become confirmable in a weak sense. However, in order to *prove* the 'normality' of the sequence we must prove an infinite number of probabilities to be equal!] Thus, even for these exceptional cases the 'limit interpretation' proves unacceptable.

[To this should be added that even in the exceptional cases just considered the virtual ensemble defined by the predicate 'P_0' would still be an *infinite class of* (infinite) *sequences*, with cardinal number larger than aleph zero. Hence the 'limit interpretation' would be *incomplete* unless extended to *all* sequences of this class. This would imply a *nondenumerable number of limit statements* as the 'meaning' of a single probabilistic law statement!]

2. As shown in Part I, the standard form of a scientific probability statement is

(II) $\text{prob}(P_0; P_1) = p_0$

where the predicates 'P_0' and 'P_1' are of *equal type* [and of the same semantic kind, as shown in Part II]. Hence, if '$P_0(x)$' is a sentence, '$P_1(x)$' is a sentence, too, and vice versa.

Let $C(P_i)$ be the *class* of all x for which the sentence '$P_i(x)$' is true, i.e. the class of all objects having the property P_i. Then the section $C(P_i) \cap C(P_k)$ is the class of all objects having the property P_i as well as the property P_k.

Now the simplest interpretation of (II) consistent with both the *syntax* of 'prob' and its *pragmatic meaning* (frequency tests) appears to be as follows:

(A)   (II) means: $\text{nc}[C(P_0) \cap C(P_1)]/\text{nc}[C(P_0)] = p_0$.

where 'nc' is an abbreviation for 'number cardinalis'.

Yet this interpretation cannot be maintained because it makes the probability statement (II) *neither confirmable nor disconfirmable*. Indeed, the interpreting sentence in (A) is a *singular* sentence (not a general law) the confirmability of which cannot be reduced to that of other sentences, and the sentence itself is neither confirmable nor disconfirmable because the number of objects having property P_0 remains unknown for any predicate 'P_0' occurring in a probabilistic law statement. [Furthermore, since $\text{nc}[C(P_0)]$ is at least aleph zero, p_0 would always be either zero or indeterminate.]

3. If we consider the tests actually used in scientific practice we see that they concern not the *class* $C(P_0)$ but *finite sets* which – as we shall say – *represent* the class $C(P_0)$. (We shall deal with the meaning of the term 'represent' below.) Some elements of such a set will have the property P_1. We may expect the number of these elements divided by the number of all

elements of the set to be nearly equal to p_0. In this way we arrive at the following interpretation:

(B)   (II) means: *in* [almost] *every finite set S large enough to represent the class* $C(P_0)$ *the number of elements of S having property* P_1, *divided by the cardinal number of S, is nearly equal to* p_0.

By an interpretation of this kind[13] probabilistic statements become *general laws*, in agreement with their *intended* meaning. [A modification of (B) is required if the predicates 'P_i' represent stochastic modes of reaction as in quantum mechanics; instead of '*having* the property P_1' we would have to write: '*acquiring* the property P_1^A (under the action of a stochastic state changer of class A)'].

If the sets S considered in (B) occur as *ordered* sets, i.e., as *sequences* [as in dice playing with *one* die] we may supplement (B) by the following axiom:

(BS)   All sequences realizing the same set S occur with [nearly] the same frequency.

Here, a sequence is said to *realize* a set iff it has the same elements, and no others. (The sequences realizing the same set differ from each other by permutations of their elements). [Sometimes the content of (BS) is expressed by saying that sequences differing merely by permutation of their elements have the same *a priori probability*; this would require an acceptable theory of *a priori* probabilities (which does not seem to exist) and its integration into the ordinary theory of probability. Presumably (BS) can be deduced from (B) and the axioms of the prob calculus by constructing a suitable prob expression descriptive of the situation that would be *invariant under the permutations concerned*. The elementary 'proofs' of (BS) or similar statements are plausible rather than strict.]

4. What is meant in (B) by '*representing*' can be seen roughly from the following example.

Let 'P_0' be the predicate 'Englishman, alive at 1st January 1939'. Then a set S_0 is said to *represent* the class $C(P_0)$ iff (i) every element of S_0 belongs to $C(P_0)$ and (ii) the composition of S_0 as to age, health, profession, and so on, of its elements is roughly the same as the composition of $C(P_0)$.

For further discussion we shall write 'repP_0(S)' for 'S represents $C(P_0)$'. (We purposely do not introduce a *two*-place predicate rep(P, S) which would allow to *define* 'repP_0(S)', because we may be able to interpret 'repP_0' for every 'P_0' occurring in a probabilistic statement of form (II) without being able to give a general interpretation for 'rep'.)

Now in order to render a probabilistic statement of form (II) *confirmable* by the interpretation (B), the statements 'repP_0(S_{oi})' must be confirmable. In general we shall meet with difficulties, indicated in the example given above by the words 'and so on', when trying to give a material definition of 'repP_0'. This is not to be considered as an objection to our interpretation (B); if we meet such a case we should regard it as a sign of imperfection of the probabilistic theory in question. Indeed, as far as the probabilistic laws of physics, genetics, or stochastic games are concerned, no difficulty as to the meaning of 'repP_0' arises.

NOTES

* Revised reprint from *Synthese, Unity of Science Forum* 3 (1938) 35–40 (Part I); 4 (1939) No. 2, 49–54 (Part II); 4 (1939) No. 4 65–71 (Part III).
[1] Cf. particularly H. Reichenbach, *Wahrscheinlichkeitslehre*, Leiden 1935; R. v. Mises, *Wahrscheinlichkeit, Statistik und Wahrheit*, Wien 1936; E. Nagel, *Principles of the Theory of Probability, International Encyclopedia of Unified Science*, Vol. I, No. 6, Chicago 1939.
[2] Cf., e.g., K. Popper, *Logik der Forschung*, Wien 1935; E. Nagel, 'The Meaning of Probability', *Journal of the American Statistical Association* 31 (1936); C. G. Hempel, 'On the Logical Form of Probability-Statements', *Synthese* (*Erkenntnis*) 7 (1938) 154.
[3] M. Strauss, 'Mathematics as Logical Syntax – A Method to formalize the Language of a Physical Theory', *Erkenntnis* 7 (1938) 147–53. [Revised reprint: Chapter VI of this volume.]
[4] R. Carnap, *Logical Syntax of Language*, London–New York 1936.
[5] H. Reichenbach, 'On Probability and Induction', *Philosophy of Science* 5 (1938) No. 1.
[6] R. v. Mises, *loc. cit.*, Note 1.
[7] H. Reichenbach, *loc. cit.*, Note 1.
[8] Following a remark by Hempel, *loc. cit.*, Note 2, the form used by Reichenbach himself is not quite correct; our form (III) corresponds to Hempel's form (II).
[9] C. G. Hempel, *loc. cit.*, Note 2.
[10] H. Reichenbach, *loc. cit.*, Note 1, p. 11.
[11] M. Strauss, *loc. cit.*, Note 3.
[12] H. Reichenbach, *loc. cit.*, Note 1, p. 11.
[13] More thorough interpretations may consider the *spread* of values p about the theoretical value p_0 in a *set of sets*.

CHAPTER IX

EQUIVALENT REPRESENTATIONS AND INEQUIVALENT INTERPRETATIONS IN PHYSICS***

The philosophical assessment, evaluation, or critique of a physical theory requires more than a mere knowledge of the theory, more than what is demanded from a physicist and what in English is called 'a good working knowledge'. For the applications of a theory it does not matter how the theory may be axiomatized, or what statements of the theory have genuine physical content and which are mere conventions. Philosophical controversies do however involve such questions from the logic of science. I need only recall here the discussions on the so-called clock or twin paradox in the Special Theory of Relativity which involve, among other things, the question of the character and role of Einstein's definition of simultaneity.

Another example, taken from the Theses prepared for this symposium, is the question whether the interpretations [of quantum mechanics] given by Fock and Blochinzew, respectively, are equivalent or not. If they are not, only one of the two can be right and it would be superfluous to consider the other in a philosophical evaluation of [present] physics. If, on the other hand, they are physically equivalent their difference is without philosophical interest.

These examples may serve to remind us that the philosophical evaluation or critique of a physical theory presupposes an answer to the scientological questions involved.

In the following, from among the wide field of the logic of science two facts are presented neglect of which has proved one of the main sources of error in philosophical discussions on science.

Our starting point is the basic fact that every physical theory (PT) consists of two parts:

1. the *mathematical formalism* M, consisting of the mathematical formalism proper [mathematical substructure] and the fundamental physical equations formulated in its language [mathematical superstructure], and
2. its *physical interpretation*, I,

in short:

$$PT = M + I. \tag{1}$$

This implies the possibility that the same physical theory may be represented by a different mathematical formalism combined with a modified physical interpretation:

$$M + I = M' + I'. \tag{2}$$

No less important is the conclusion that a changed interpretation of the same formalism will yield a different physical theory:

$$M + I \neq M + I'. \tag{3}$$

Some examples will illustrate these conclusions.

Some people have asked whether the so-called clock or twin paradox of Special Relativity may not result from a clumsy choice of units or from a [clumsy] definition of simultaneity. This question has been investigated mathematically in recent years, by generalizing the Lorentz transformation in two different ways. In one generalization[1] the *units* of length and time or – what amounts to the same – of length and velocity are allowed to differ for different inertial frames [so that new parameters appear in the formalism]. The other generalization[2] results from admitting different [non-Einsteinian] definitions of *simultaneity* for different space directions, subject to the condition

$$1/c_x + 1/c_{-x} = 2/c$$

for *opposite* directions as required by the Michelson and similar experiments, whereby the propagation of light becomes [formally] anisotropic.

The two mathematical formalisms thus obtained differ from the old one but when suitably interpreted they are completely equivalent to the old familiar theory.[3] It should be noted that neither of the new formalisms is isomorphic with the old one; since they possess a richer structure they could even be used to represent a richer field of experience. Thus, *the [rather frequent] view that mathematical formalisms have to be isomorphic if they are to represent the same theory is untenable.*

Vice versa, *isomorphic formalisms can represent different physical theories.* I recall the well-known isomorphism between the mathematical theories of *geometrical optics* and *classical mechanics* which may both be deduced from the same variational principle.

To illustrate the second scientological fact mentioned above, viz., the fact that *different interpretations of the same formalism lead to inequivalent physical theories*, we draw on [the discussions of] quantum mechanics. As well-known, a so-called *causal interpretation* of the quantum mechanical formalism [or rather of its Schroedinger representation] was advanced by Bohm who believed this interpretation to be physically equivalent to the usual interpretation. This belief has been shown to be wrong. E.g., according to Bohm's interpretation an atomic hydrogen gas [with all atoms in the ground state] ought to be *paramagnetic* while it is predicted to be *diamagnetic* by the standard interpretation.

[Incidentally,] several other quite different interpretations of the quantum mechanical formalism are possible some of which are even suggested by the formalism itself. Thus, it would be quite in line with the formalism if the state of a [quantum physical] system after an unsharp measurement [or 'preparation'] would be represented not by a *direction* in Hilbert space but by a *many*-dimensional subspace (or the corresponding projection operator) [with the number of dimensions corresponding to the unsharpness]. Such a physical interpretation [of the subspaces of Hilbert space] would yield a logically consistent but physically wrong theory.

As for the physical interpretations advanced by Fock on the one hand and by Blochinzew on the other hand, I would like to say this. It is of course possible, as suggested by Blochinzew, to coordinate to the state vector representing the state of a *single* system a *whole ensemble* of such systems all in the same state. This may simplify the formulation of the probabilistic interpretation of the state vector [in terms of ordinary statistics] but taken by itself it does not change the physical content of the theory in the least. However, Blochinzew has allowed himself to be misled by his reformulation to make an assertion that is *inconsistent* with the standard interpretation. He maintains that it is *impossible* to ascertain the state of a system otherwise than by ascertaining its membership in the ensemble. This is not only wrong but complete nonsense since it would make the [experimental] definition of 'state' circular and empty. [By the way, the ensemble that *can* be coordinated to the state vector is of course a *virtual* ensemble while Blochinzew speaks of it as if it were a *real* ensemble.] It follows that the characterization of the difference between the two interpretatons given in the Theses, no matter whether it is right or wrong, misses the essential point.

Of course, much more can be said on the theme of this contribution. Yet my sole aim was to present two fundamental facts from the Logic of Science neglect of which leads to wrong inferences and quite unnecessary discussions.

NOTES

* Translated from 'Über äquivalente Darstellungen und nichtäquivalente Interpretationen in der Physik', *Wissenschaftliche Zeitschrift der Humboldt-Universität zu Berlin* **14** (1965) 585-6.
** Based on the author's contribution to the 1964 Symposium on 'Gesetz und Bedingung in den technischen und Naturwissenschaften' organized by the Section 'Philosophical Problems of Modern Natural Sciences' of the Institute for Philosophy, Humboldt University at Berlin, 21-22, September, 1964.
[1] Cf. J. E. Romain 'On Some Misconceptions About Relativistic Co-ordinate Transformations', *Nuovo Cim.* **30** (1963) 1254-71.
   The mathematical structure of Romain's generalized Lorentz transformation and its possible physical interpretations have been investigated in: M. Strauss, 'On a Generalized Lorentz Transformation', *Ann. d. Phys.* (7) **16** (1965) 105-13.
[2] Cf. W. F. Edwards, 'Special Relativity in Anisotropic Space', *Amer. J. Phys.* **31** (1963) 482-9, cf. also Note 3.
[3] For the second generalization this holds only for light kinematics but not for a combined kinematics for light and particles, to say nothing about full mechanics, contrary to what has been suggested by Edwards. For a discussion of these points, with refutation of Edwards' conventionalistic position, cf. M. Strauss, 'The Lorentz Group: Axiomatics–Generalizations Alternatives', reprinted as Chapter XIV of this volume.

CHAPTER X

MATERIAL STRUCTURE AND MATHEMATICAL STRUCTURE***

My theme is the relation between *material structures* and *mathematical structures*. In the 'Theses' composed for this Symposium definitions are given of what is to be understood by 'mathematical structure' and 'material structure' (there called 'structure in the physical object region'). Yet no statement can be found concerning the relation between the two. True, Thesis Seven says that when the physical content is disregarded *physical structures* are identical to mathematical ones; however, we are left guessing what is meant by 'physical structure'. If *material* structure is meant the thesis is wrong if not nonsense. If the structure of a *physical theory* is meant the thesis is true in a trivial sense and does not say anything about the relation between material and mathematical structure.

This is not meant as a reproach. The clarification of the relation between the two kinds of structure does not fall into the competence of either mathematics or philosophy, though it is of interest to both; it falls into the competence of theoretical physics which uses mathematical structures for the characterization of material structures.

As a typical example of a *real* structure let us consider a conservative system consisting of N mass points together with their interactions. This (idealized) material structure is represented both in classical and in quantum mechanics by a Hamilton function $H(q_i, p_i)$ of $2 \times 3N$ canonical variables $q_i, p_i (i = 1, \ldots 3N)$. However, there is an infinity of possible choices of canonical variables[1], and hence an infinity of possible Hamilton functions, all different from one another. Thus *to one and the same material structure belongs an infinity of mathematical structures*.

This result, which is easily generalized to cover all parts of physics, is the death warrant for any naturalistic (abstraction theoretical) conception of mathematics. The basic mistake inherent in this conception is the failure to realize that it is only the *union* of a mathematical structure and its physical interpretation that has an invariant physical meaning. Hence the *logic of physics* can be reduced neither to mathematical logic nor to pure semantics [or pragmatics] as the operationalists are trying to do. The logic of physics is syntacto-semantics, i.e., the union of syntax and semantics. Indeed, all non-trivial problems in the logic of physics turn out to be of a syntacto-semantical nature! To mention some typical examples: What is the *number of independent physical dimensions* for a given theory? Which of the proposed *'logics' for quantum theory* is correct or acceptable? Another example is the question of *physical geometry*. It would not be a meaningful question to ask what the metrical structure of *real space* is since physical geometry depends on the physical interpretation ['coordinating definitions'] of the mathematical concepts. This follows already from the well-known considerations of [Riemann, Helmholtz,] Poincaré and Reichenbach. Today, we know for instance that on a rotating disc Euclidean geometry can be maintained for solid bodies while 'light geometry' is definitely non-Euclidean.

Thus, the mathematical structure [of a physical theory] is uniquely fixed only if the coordinating definitions are given. This applies also to the example discussed above. Indeed, if the q_i are to be interpreted, e.g., as the *Cartesian* coordinates of the mass points the Hamiltonian of the system is uniquely determined.

For most physicists all this is trivial. It would be nice if mathematicians and philosophers, too, would take cognizance of these facts and would correct their often much too primitive conceptions concerning the relation between mathematics and physical reality.

Finally, I present an example to show that one and the same mathematical structure may represent two physical theories that are not only *different* [such as geometrical optics and point mechanics] but even *incompatible*.

Let M be the set of all real numbers and M' the set of all real numbers between $-c$ and $+c (c > 0)$. Now the elements a of M form an Abelian group with respect to ordinary addition. Likewise the elements a' of M' form an Abelian group with respect to the composition

$$a' \dotplus b' =_{df} \frac{a' + b'}{1 + a'b'/c^2}. \tag{1}$$

Moreover, the two sets M and M' have the same cardinal number. Thus, the two structures $(M, +)$ and $(M', \dotplus)$ are *isomorphic*. The one-one-mapping implied is given by the function

$$a' = c \, \text{th} \, (a/c), \quad (\text{th} = \text{tangens hyperbolicus}). \tag{2}$$

Nothing is changed if we interpret a and a' as the values (real number times unit) of collinear velocities; in M', c is then the limiting velocity and the composition law (1) given above becomes identical with the Einstein law for the composition of collinear velocities. Thus, the same algebraic structure represents both the Galilean and the Einstein law. The difference between the two laws lies solely in the different velocity *metrics* and does not affect the algebraic structure. This difference would even disappear if the velocity *scale* [in one or the other theory] would be changed (regauged) according to (2).

Such a regauging, to be sure, is not permitted in our case since it would change the topological and algebraic structure of *other* parts of the theory concerned. This is an instructive example of the rule that [in general] the *scale metric* is *not* a mere convention, and thus a further argument against operationalism.

However, the main lesson to be learned from our example is the recognition that *mathematical isomorphism is not the same as physical equivalence*.

Summarizing, it may be said that there is no unique relation in either direction between *mathematical* structures and *material* structures. This underlines our previous remark that for the purposes of real science mathematical logic is quite insufficient and has to be replaced by general logic, i.e., by syntacto-semantics.

NOTES

* Translated from 'Über das Verhältnis von Realstruktur und mathematischer Struktur', *Wiss. Z. Humboldt-Univ. Berlin, Math.-Nat. R.* **16** (1967) 871-2.

** Paper read at the Symposium 'Struktur und Prozess in den Naturwissenschaften und der Technik' organized by the Section *Philosophische Probleme der modernen Naturwissenschaften* of the *Institut für Philosophie* of the Humboldt University at Berlin and held on 21-22 November, 1966.

[1] If (q_i, p_i) are canonical variables, so are all variables (Q_k, P_k) resulting from the former by canonical transformation, i.e. a transformation that does not change the Hamiltonian form of the law of motion.

CHAPTER XI

SPACE-TIMES AND STATE SPACES*

An exhaustive treatment of this theme is impossible in this discussion. I shall therefore confine myself to some remarks of philosophical relevance.

1. The general notion 'state Z of physical system S at time t', abbreviated $Z(S, t)$, may be [axiomatically] defined as follows:

$Z(S, t)$ determines the probabilities of the possible responses of S to external influences starting at t; if no external influence is acting on S the state of S changes according to a *law of motion* of the form

$$Z(S, t + \tau) = \Omega(S, \tau) Z(S, t) \tag{1}$$

with some operator $\Omega(S, \tau)$. The set M_S of all possible $Z(S, t)$, for fixed S, is the *state space* of S; the operator $\Omega(S, \tau)$ is the *motion operator* [or *propagator*] for S.

From (1) it follows that

$$\Omega(S, 0) = I \quad \text{(identity operator)} \tag{2a}$$

and

$$\Omega(S, \tau_1 + \tau_2) = \Omega(S, \tau_1) \Omega(S, \tau_2) = \Omega(S, \tau_2) \Omega(S, \tau_2) \tag{2b}$$

[Thus, the motion operators $\Omega(S, \tau)$, for fixed S, form a 1-parameter Lie group with $\Omega(S, 0)$ as unit element and the *time interval* τ as parameter].

Of philosophical relevance is the fact that Equations (2) *determine the time metric* [equality of time intervals] *uniquely*, [and hence the *time scale* up to an arbitrary factor and an arbitrary addititive constant]. Any time [scale] with this metric is called *canonical time* [scale].

2. The axiomatic definition given above is applicable to closed non-relativistic systems only; it may however be generalized so as to cover relativistic fields.

3. Another possible generalization would consist of replacing the motion operator $\Omega(S, \tau)$ by a stochastic operator correlating to $Z(S, t)$

a *set* of states $Z_i(S, t+\tau)$ with probabilities $p_i(\sum p_i = 1)$. (A theory of this kind has first been advanced by REICHENBACH[1]). However, [fundamental] physics has made no use of this possibility so far.

4. According to the above axiomatic definition the notions of *state* and *motion* are interrelated: 'state' has always to be defined in such a way that the motion of the system is describable as a mapping of the state space onto itself. On the other hand, if the state space is known the equation of motion is partly determined by (1): since the general solution of Equation (2) reads

$$\Omega(S, \tau) = \exp(\tau B_S), \tag{3}$$

only the τ-independent operator B_S has to be found. In the case of quantum mechanics we have

$$B_S = iH_S/h \tag{4}$$

where H_S is the Hamilton operator for S. In the case of classical mechanics we have[2]

$$B_S = \sum_i \left[\frac{\partial H}{\partial p_i} \frac{\partial}{\partial q_i} - \frac{\partial H}{\partial q_i} \frac{\partial}{\partial p_i} \right], \qquad H = H_S. \tag{5}$$

In the theory of fundamental particles it would signify a big advance if we were to know the [mathematical structure of the] state space.

5. The mathematical structure of the state space mirrors essential properties of the system concerned as everybody can work out for himself.

6. The fundamental role of the state concept may, e.g., be seen from the fact that the concept of a *quantum mechanical quantity* ['observable'] can be defined in terms of the state concept.[3] A *quantum mechanical* [retrospective] *measurement* may then be defined as an external influence that compels the system to go momentarily into one of the eigenstates of the measured quantity ('momentarily' because *after* the measurement the state may again be different or – in the case of absorption – nonexistent).

7. If two physical systems are to be treated as a single *composite system* the state space of the latter results from the state spaces of the component systems (a) as their *direct product*[4] in the case of classical mechanics, and (b) as their *direct sum*[4] in the case of quantum mechanics. This difference in structure is of great physical import: it leads to what SCHROEDINGER[5] has called 'Verschraenkung' [a particular

instance of which is the much discussed thought-experiment of
EINSTEIN-PODOLSKY-ROSEN [6]].

8. The space of the (classical) q_i is called *configuration space*; for a
single mass point it becomes identical with the ordinary Newtonian space.
By contrast, the state space of quantum mechanical systems is always a
Hilbert space of infinitely many dimensions and a configuration space
does not exist.

9. The kinematic transformations (of the Galileo or Lorentz group)
induce corresponding transformations in the state space, called *representations* of the group concerned in the state space. Now the kinematic
transformations themselves are representations of the algebraic velocity
group in space-time.[7] Hence, the corresponding transformations in state
space can likewise be considered as representations of the algebraic
velocity group. From this point of view *state space and space-time are
alternatives with no claim to logical priority or superiority with respect to
one another*. Furthermore, there is, strictly speaking, no space-time
in quantum mechanics in the sense of a space of events [the quantum
mechanical events like state transitions being events in the state space].
*Thus the concept of state space appears not only more general but also
more fundamental than the concept of space-time.*

The position operators of quantum mechanics may be considered as
remnants of space-time. In quantum field theory classical Minkowski
space-time reappears as region of definition of the field operators, which
is one of the main sources of trouble in this theory. Hence a future theory
of fundamental particles may be expected not to contain any space-time
at all in its mathematical part but only in its physical interpretation.

10. The theoretical devaluation of space-time brought about by quantum theory is in marked contrast to Einstein's theory of gravitation and
certain trends based on it such as geometrodynamics. This leads to the
following remarks. [The remarks 11–13 repeat points made in other
papers contained in this volume and are here left out.]

14. Einstein's theory is as much a *physicalization of space-time* as a
geometrization of gravitation. In particular, space-time is now endowed
with a [metrical] *structure* of a kind allowing, in principle, identification
of any of its elements [i.e. points] [namely by the value of the curvature
tensor in that point]. Hence one may be tempted to regard Einstein's
space-time as an object of independent physical significance [independent,

to wit, of physical matter in the ordinary sense as represented by the matter tensor T_{ik} in Einstein's equations].

15. In order to test the consequences of such a conception one has to put the matter tensor equal to zero and to investigate [the homogeneous Einstein equations and] the *Einstein spaces* (in the narrower meaning of this term) defined thereby. If one now introduces 3-dimensional subspaces $t =$ const. as time-dependent physico-geometrical objects one arrives indeed at a *geometrodynamics*. It is WHEELER's[9] incontestable merit to have carried out this program to a large extent. However, even in Wheeler's opinion it is quite uncertain whether the mathematical results thus obtained have any physical significance. In my view geometrodynamics ascribes to the mathematical formalism of Einstein's theory an ontological significance that it does not possess. As is well known from Maxwell's theory, satisfying the field equations is a *necessary* but *not* a *sufficient* condition for the solution to represent physical reality. [Apart from this general observation a specific objection can be raised against geometrodynamics]. Only in the case of *linear* equations can we ascribe a definite physical meaning to the solutions of the *homogeneous* equations (source term $= 0$) because of the *superposition principle* valid for such equations and because the *general* solution of the original *inhomogeneous* equations can be written as the *sum* of the general solution of the *homogeneous* equations and a *particular* solution of the *inhomogeneous* equations. [In contrast to Maxwell's equations] Einstein's equations are highly *nonlinear* and hence the *homogeneous* Einstein equations ($T_{ik} = 0$) *cannot* be assumed to have any physical significance, except as a limiting case for small curvature. [Thus, Wheeler's heuristic analogy with the Maxwell theory is without mathematical foundation].

16. The results of our analysis suggest that the *theory of fundamental particles* and *Einstein's theory of gravitation* may be *complementary* to each other in the sense that there exist no elementary processes in which *both* space curvature and strong interactions play a significant role. Whether this surmise proves correct only future development can show.

NOTES

* Translated from 'Struktur der Zustandsräume und der Raum-Zeit', *Wiss. Z. Humboldt-Univ. Berlin, Math.-Nat. R.* **16** (1967) 987–8.
Discussion remarks made in the Section Physics/Chemistry of the Symposium 'Struktur

und Prozess in den Naturwissenschaften und der Technik' organized by the Section *Philosophische Probleme der modernen Naturwissenschaften* of the *Institut für Philosophie* of the Humboldt University at Berlin and held on November 21–22, 1966.

[1] H. Reichenbach, 'Stetige Wahrscheinlichkeitsfolgen', *Z. Phys.* **53** (1929) 274. [Another mathematical theory based on the same idea and successfully applied is that of *MARKOFF chains.*]

[2] Cf. e.g. H. Steudel, 'Symmetrieeigenschaften und Erhaltungssätze', *Diss.* (1966) Humboldt University Berlin (preprint). [The mathematical theory involved in writing the classical law of motion in this form is the theory of *LIE series*].

[3] Cf. M. Strauss, 'Grundlagen der modernen Physik' in *Mikrokosmos-Makrokosmos* (ed. by H. Ley and R. Löther), Berlin 1967 [Chapter XIX of this volume]; 'Quantentheorie und Philosophie' in *Naturwissenschaft und Philosophie* (ed. by G. Harig), Berlin 1960 [Chapter XX of this volume.]

[4] [This corrects a careless mistake in the original where 'direct sum' and 'direct product' have changed places.]

[5] [E. Schrödinger, 'Die gegenwärtige Situation in der Quanten-mechanik', *Naturwiss.* **23** (1935) pp. 807, 823, 844.]

[6] [A. Einstein, B. Podolsky, and N. Rosen, 'Can Quantum Mechanical Description of Physical Reality be Considered Complete?', *Phys. Rev.* **48** (1935) 777.]

[7] M. Strauss, 'The Lorentz Group: Axiomatics – Generalizations – Alternatives', *Wiss. Z. Friedrich-Schiller-Univ. Jena, Math.-Nat. R.* **15** (1966) 109 [Chapter XIV of this volume].

[8] [Cf. Chapters XIV and XV of this volume.]

[9] J. A. Wheeler, *Geometrodynamics*, New York 1962; 'The Universe in the Light of General Relativity' in *Lectures in Theoretical Physics*, Vol. V, (ed. by W. E. Brittin, B. W. Downs, and J. Downs), New York – London – Sydney 1963.

CHAPTER XII

INTERTHEORY RELATIONS I –
GENERAL PROBLEMS*

INTRODUCTION

While there exists a vast literature devoted to logical analysis of physical theories the literature on intertheory relations is almost nonexistent. True, intertheory relations are touched upon by many authors, both scientists and philosophers of science, and the interest in intertheory relations is rising rapidly. Yet the systematic study of them has hardly begun, a notable exception being the work of TISZA [1] which, however, is complementary to rather than concurrent with the present work.

Thus I shall have to break new ground, and in doing so I may pose more queries than I can answer, and some of the answers may turn out to be wrong. This is a risk everybody who ventures into new fields of study has to take. To minimize the risks it may be wise to reflect in advance on some of the issues, methodological and other ones, that are involved in a study of ITRs'. These reflections are offered as a kind of provisional substitute for a still nonexistent ITR theory and it will be best to start the reflections with a list of desiderata that a general ITR theory should satisfy.

1. Desiderata for a General ITR Theory

A general ITR theory would have to supply a *well-organized system of syntacto-semantic relational concepts in terms of which all possible ITRs could be formulated.* The fundamental problem to be solved would be the construction of a *model$_1$ of the logical structure of physical theories*: this model must be (1) *realistic*, i.e., reflecting the real (and not an imaginary) logical structure of physical theories, (2) *sufficiently general* so as to encompass all physical theories, and (3) *sufficiently detailed* to allow a sufficient characterization of any given physical theory.

2. Realistic Model of Logical Structure

I shall not discuss desiderata (2) and (3) any further since a perusal of the relevant literature shows that even the first one is rarely satisfied and often grossly violated. The most fashionable model$_1$ of the logical structure of physical theories seems to be the Duhem-Quine model ('hypothetic-deductive system') – a model that has been taken over from *Mathematical Logic*[2] (where it is alright) to Physics where it is entirely unrealistic. From the *logical point of view* a physical theory is not a hypothetic-deductive but a *syntacto-semantic system with pragmatic import*, or, in ordinary language, a *mathematical formalism* (MF) *together with a physical interpretation* (PI). The MF, whether taken by itself or in conjunction with PI, does not supply a set of 'implicit definitions' (as a hypothetic-deductive system is supposed to do) but determines the *syntax* of the theoretical concepts as is most clearly seen when the theoretical language is formalized[3]; alone, it says nothing at all about the (semantic) meaning of the mathematical symbols used. Hence the whole *doctrine* of 'implicit definitions' 'meaning relative to the system' etc. is untenable and has to be replaced by the *question* to *what extent* a change in syntax implies, or rather *reflects*, a *change in meaning* – a question that cannot be answered in a general way but requires detailed investigation in any given case.

Thus *variance of meaning* of a theoretical term under transition from T_1 to T_2 is a legitimate problem of ITR-theory, but it has nothing to do with the doctrine mentioned above. In most cases the problem can be resolved by distinguishing between *nuclear* or *core meaning* and *peripherial* or *subsidiary meaning*. This applies for instance to 'velocity$_1$' (= 'velocity' as used in Newtonian Kinematics) and 'velocity$_2$' (= 'velocity' as used in c-Kinematics) where the difference in syntax does not reflect any essential difference in meaning and none at all of pragmatic import.

3. *Variance of Meaning*

If there is an essential change in meaning under transition from T_1 to T_2, as is the case with almost all 'corresponding terms' of o-PM and h-PM, we should use a *differentiating terminology* indicating both the common core (if any) and the difference in peripherial meaning. If we do not do so, confusion is bound to arise. A case in point is the use of the undifferentiated term 'probabilistic' or 'statistical' as in 'statistical interpretation of QM'. The probabilities of QM refer to *stochastic transi-*

tions between states induced by interaction with macroscopic systems, and not to a distribution of properties in an ensemble. No wonder that the *mathematical syntax* of these transition probabilities is *essentially different* from that of distribution probabilities as given in axiomatic form by Reichenbach and Kolmogoroff: while the field of definition of $prob_1$ is the set of ordered pairs of elements taken from a *Boolean* algebra, the field of definition of $prob_2$ is the set of ordered pairs of elements taken from a *semi-Boolean* algebra, the latter being represented in QM by projection operators.[4]

4. *Logical Analysis and ITR Study: New Concepts Required for Non-Trivial ITRs*

I now turn to the question: *what is the relation between logical analysis*, viz., the study of the syntax and semantics of a physical theory, *and the study of ITRs?*

The answer is twofold. On the one hand, a great deal can be said about ITRs that does not depend on a detailed logical analysis of the theories compared. Thus, in some respect intertheory relations say less than can be said by logical analysis, for about the same reason that the equation $f(x)/g(x)=x^2$ says less than the two equations $f(x)=a\,x^5$ and $g(x)=a\,x^3$.

On the other hand, the study of intertheory relations transgresses logical analysis in a fundamental way, and this for two reasons. First, to state non-trivial intertheory relations we need *new relational terms not used in ordinary logical analysis*. Since a physical theory is a syntacto-semantic system, these new relational terms will, in general, refer to both syntax and meaning, i.e., they will be *syntacto-semantic terms*.

The second reason is this. In ordinary logical analysis semantics can be *technically reduced* to syntax by a device best known from Carnap's reduction sentences, viz., by using a *mixed language* containing terms belonging to both the interpreting language and the language to be interpreted. Even if, instead, a common meta-language is used to state the interpretation (e.g., in form of 'truth conditions') the situation does not change fundamentally, viz., semantics is still *effectively reduced* to syntax (in the meta-language). *Such a technical or effective reduction of semantics to syntax does not seem possible in the study of ITRs*, as the above discussion on 'variance of meaning' indicates.

5. *Dialektische Aufhebung and 'Dominance'*

The difficulties just outlined can only be overcome by introducing *new syntacto-semantic relational terms appropriate to state non-trivial ITRs*. Some such terms do indeed exist but they have been invented by dialecticians and not by formal logicians. The most important of these terms is *dialektische Aufhebung* [uplation, dialectic negation]. It correctly describes all of the following ITRs: c-K-theory to c-theory, c-mechanics to o-mechanics, h-mechanics to o-mechanics, statistical thermostatics to macroscopic thermostatics. Yet it does *not* describe the *differences* between these ITRs which are equally important; neither does the even more general term 'dominant over', used by *Tisza* for these and other relations. It would be wrong to discard these general terms altogether – that would mean *den Wald vor lauter Bäumen nicht zu sehen* – but it would be equally wrong to renounce a more specific description of ITRs. Indeed, the trouble with philosophical concepts is not that they are semantically ill-defined – this they share with all theoretical concepts – but that they are *too general*. They have to be split up into a spectrum of more specific concepts.

6. *Splitting up of Degenerate Concepts*

A more specific but still fairly general term that will be used repeatedly in ITR study is *conceptual degeneracy* (in T_1 with respect to T_2) or, for the inverse relation, *conceptual differentiation* or *concept splitting*. A concept of T_1 is said to be *degenerate in T_1 with respect to T_2* if it splits up into two (or more) concepts in T_2 or, equivalently, if two (or more) concepts of T_2 have the *same extension* in T_1. A typical example is '(inert) mass' which splits up into 'rest mass' and 'relative mass', the latter being the correct measure of inertia in c-mechanics as the case of photons shows.

Another important and still fairly general ITR term, introduced by the writer, is *partial anticipation of T_2 in* (or *by*) T_1. I apply this term to *ad hoc* postulates in T_1 that become *deducible* in T_2. An instance is the postuale of QM connecting spin value and permutational parity of state vectors; it is deducible in h-c-theory.

7. *Limit Relation between Ranges of Validity*

An important semantic concept that belongs to ordinary logical analysis but is of particular import for ITR studies is RD(T), meaning *'region of definition of T'*, in contradistinction to RV(T), meaning *'range of (approximate) validity of T'*, with the obvious relation

$$RV(T) \subseteq RD(T). \tag{1}$$

As is well known the concept of *temperature* presupposes a state of thermodynamic equilibrium and hence thermostatics cannot be applied to non-equilibrium systems. Even the (present) thermodynamics of irreversible processes is essentially limited to transport and similar processes where the temperature depends but weakly on time and space; hence it cannot be applied to turbulent systems. In all these cases the limitations in applicability are not due to a breakdown of the theory but to the fact that the *preconditions underlying the theoretical concepts* are not fulfilled: they determine the RD of the theory.

   Now the comparison of theories is of little or no interest unless their RDs are equal or, at least, overlapping. In case they are equal, $RD(T_1) = RD(T_2)$, let $RV(T_1)$ be contained in $RV(T_2)$: $RV(T_1) \subset RV(T_2)$. We can then ask: under what conditions $\{C\}$, if any, imposed on T_2 does $RV(T_2)$ contract to $RV(T_1)$? If there are such conditions we write

$$RV(T_2) \underset{\{C\}}{\rightarrow} RV(T_1). \tag{2}$$

A well-known example is the pair 'geometrical optics' (T_1) and 'wave optics' (T_2). The conditions C reads: all $\lambda \ll$ all d, d being any of the relevant linear dimensions.

8. *Other Types of Limit Relations*

The question answered by the scheme (2) is but one of several ITR *transition* or *limit problems* that should be clearly distinguished.

   In the first place, we have to distinguish in the usual way between *asymptotic limit* ($\rightarrow$) and *exact limit* (lim). This mathematical distinction is, in itself, not of great logical import, but it may become so in connection with semantics.

   In the second place, we must distinguish between *three kinds of conditions* C:

(a) those that can be formulated *within* T_2,

(b) those that can be formulated in the *meta-language* of T_2,

(c) those that require reference to *extratheoretical objects*, or applications, as in the example above.

To see the difference between (a) and (b), consider the operator

$$\lim_{v/c\,=\,0}$$

which may be interpreted as either

$$(\alpha)\ \lim_{\substack{v\to 0 \\ c\,=\,\text{const}}}$$

or

$$(\beta)\ \lim_{\substack{c\to\infty \\ v\,=\,\text{const}}}$$

If c is the universal constant of a c-theory T_2, the operator (α) is defined *within* T_2 while the use of the operator (β) is forbidden in T_2 where c is a constant. However, we may treat c as a variable parameter in the meta-theory of T_2.

If T_1 is o-PM and T_2 c-PM we have:

(A) *no*[5] limit relation between the space-times of T_1 and T_2,

(B) *two* limit relations between the two *kinematics* (transformation formulae etc.):

$$\lim_{c\,=\,\infty} T_2^{K} = T_1^{K} \tag{3}$$

$$T_2^{K} \xrightarrow[(v/c\,\ll\,1)]{} T_1^{K}. \tag{4}$$

(C) *no* limit relation of *general* validity between the two *dynamics* because of $E^2 - c^2 P^2 = m^2 c^4$, but again *two* limit relations between corresponding *concepts* (momentum $\mathbf{P}$ and kinetic energy $E^{\text{kin}} = E - mc^2$):

$$\lim_{c\,=\,\infty} \mathbf{P}_2 = \mathbf{P}_1, \qquad \lim_{c\,=\,\infty} E_2^{\text{kin}} = E_1^{\text{kin}} \tag{5}$$

$$\mathbf{P}_2 \xrightarrow[(v/c\,\ll\,1)]{} \mathbf{P}_1, \qquad E_2^{\text{kin}} \xrightarrow[(v/c\,\ll\,1)]{} E_1^{\text{kin}}. \tag{6}$$

Thus o-kinematics and o-dynamics are both (a) definable *within* c-mechanics as *asymptotic* limits and (b) definable in the meta-language as *exact* limits, although there is *no*[5] limit relation between the underlying space-times.

This example shows the importance of *differentiating between the different layers or levels of a theory* when limit relations are considered.

9. *Equivalent Formulations Cease to be Equivalent*

While the problems mentioned so far have been discussed by other authors the following problem does not seem to have been noticed at all: *why does a physical theory admit of different mathematical* formulations all *normal* in the sense that they do not contain any redundant parameters[6]? In particular, why does o-PM admit such different formulations as those connected with the names of Newton, Lagrange, Hamilton, Hamilton-Jacobi, Poisson, which are all normal in the sense explained.

The first answer one is tempted to give is this: all these formulations are mathematically equivalent and hence no genuine problem is involved in the multitude of formulations.

This answer is *deceptive* for reasons that are best explained by analysing the ITRs between Newtonian theory (T_3), the Copernican 'system' (T_2), and the Ptolemaic 'system' (T_1). Since T_1 and T_2 are kinematic descriptions based on the same space-time theory, and since, in this space-time, all possible frames are kinematically equivalent, T_1 and T_2 are but *equivalent formulations* of the *same* 'theory'[7], mathematically connected by one of the 'admissible' transformations (full space-time group). In spite of this, *the two formulations T_1 and T_2 cease to be equivalent when judged from the standpoint of T_3*, as the transformation connecting T_1 and T_2 does not belong to the covariance group of T_3 (Galileo-group), which is but a small subgroup of the full space-time group.

Thus, by analogy, we may expect that *the different formulations of o-PM cease to be equivalent* when judged from the standpoint of either c-PM or h-PM. This is indeed the case and the reasons are again to be found in the differences between the covariance groups characteristic for o-PM, h-PM and c-PM, respectively, though the ITRs are here somewhat more involved since the different formulations of o-PM differ themselves in their respective covariance groups. This will be more fully explained in Part III. If one of a set of equivalent formulations of T_1 is singled out as preferential from the standpoint of T_2, we may say that it is a *partial formal anticipation* of T_2.

More often than not, the singled-out formulation is the one of maximal formal simplicity: thus *formal simplicity may have heuristic value*.

10. *Three Types of Physical Axiomatics*

So far I have said nothing on axiomatic formulations of physical theories, for the simple reason that I shall not need them. We may ask, however, whether such formulations can help in ITR study. To answer this question we must distinguish three main types of physical axiomatics which I shall call 'constructive', 'ordinary' and 'deductive'.

Constructive axiomatics aims at reconstructing the MF *together* with its PI from *low-level postulates* that do not require physical interpretation and can be tested separately.

Ordinary axiomatics is the attempt to apply the axiomatic method as used in mathematics to physical theories. It takes the MF for granted and supplements it by semantic axioms. Thus, in effect if not in method, it axiomatizes the physical interpretation.

Deductive axiomatics starts with *high-level* axioms ('principles') that allow either to *deduce* the MF or to show the latter to be a *model$_2$* of the axioms. Thus, there are *two versions* of deductive axiomatics, to be called *concrete* and *abstract*, respectively. Examples of concrete deductive axiomatics are: Einstein's 1905 paper founding *c*-theory, Caratheodory's restatement of thermostatics. Typical examples of abstract deductive axiomatics are LANDÉ's [8] reconstruction of QM, the present author's 1936 paper on QM [9] and the more recent work on *c*-kinematics (transformation groups as representations of abstract velocity group) [10].

Now it should be obvious that each of the three axiomatic methods has its merits and its drawbacks or limitiations which I shall not discuss here. What I *will* say is this:

First, *ordinary* axiomatics sheds *no light* on the most important question, viz., on the question why a particular MF$_s$ rather than another one should be used to formulate the physical laws.

Second, *constructive* and *deductive* axiomatics *do just this*, but they approach this aim from opposite directions; thus they are complementary.

Third, for ITR study the *abstract version of deductive axiomatics* is the most useful one: it offers the possibility of a *common axiomatic basis* for two (or more) theories which are different models$_2$ of the same high-level structure; the differentiation between these theories can then be achieved by *specific branching* axioms.

11. *Prehistory of the Subject and Recent Work*

Next, a few words should be said here on the prehistory of our subject and some recent work in this field.

At the time of Galileo, the Copernican and the Ptolemaic 'systems' were considered as rival theories, not as equivalent descriptions differing only in descriptive simplicity, and this *then* mistaken view prevailed for centuries, in some quarters until today. *Rival or competing* theories, i.e., different theories with the same RD, have become the play ground for some logicians preoccupied with questions of confirmation, 'probability of theories', etc. But mere rivalry or competition is not a characteristic relation between any two modern theories though it cannot be denied that it has played a great role in the history of physics.

The first instance of two full-fletched theories standing in the relation of *dialektischer Aufhebung* characteristic for the *modern* development of physics is no doubt given by macroscopic and statistical thermostatics (STS), and this is probably also the most widely discussed instance – one reason why it will not be discussed in the following pages although the exact nature of the ITRs between the two theories is still a matter of controversy [11]. But there are three queries I cannot suppress.

First, there exist again several equivalent, or nearly equivalent, formulations of STS and one would like to known which of them, if any, is singled out as preferential from the standpoint of statistical thermo-*dynamics* (irreversible processes).

Second, the ITRs between STS and *o*-mechanics appear to be rather different for different formulations of STS and hence one may ask: which formulation is singled out as preferential by *h*-mechanics? If it should turn out that the *h*-preferential formulation is not the preferential formulation of the previous query, why?

Last but not least, what becomes of the beautiful work of Einstein on fluctuation theory if the Boltzmann-Planck formulation ($S = k \ln W$) is replaced by the Gibbs formulation?

A new idea entered ITR study in connection with *c*-mechanics. After Einstein had found *c*-dynamics by the *detour* over electrodynamics, Tolman noticed that it could be obtained more directly by postulating

Lorentz-invariance of conservation laws (applied to collision) and the *limit relations* (5), (6) given above.

The existence of limit relations may be said to be the core of Bohr's *correspondence-principle*. However, if this 'principle' is analysed with due regard to its context it will be found to contain several components of a rather novel character. First, *the limit relation* is not demanded for any level or layer of the two MFs but only for the *level* of *application*. Second, the limit is demanded not for $h \to 0$, which would correspond to $c \to \infty$, but for $n \to \infty$, n being a *quantum number*. Third, the principle provides a sort of implicit definition for 'corresponding concepts'. To be sure, Bohr's correspondence principle was conceived as a *heuristic* principle both for finding a new mathematical formalism and for finding the correct physical interpretation of it. But it still is an inherent part of h-theory as understood to-day, in marked contrast to other heuristic 'principles' that have not outlived the grown-up theory they were supposed to found. In this, Bohr's correspondence principle seems to be unique.

After I had written the greater part of this paper I discovered TISZA's [1] paper of 1963. As I have said before, this paper is complementary to rather than concurrent with the present one, and this applies to both the theories selected for more detailed ITR study (thermodynamics and electrodynamics being preferred) and to the spirit of approach (empiricist rather than realist). However, there are many 'points of contact' between the two papers and at these points agreement is about as often as disagreement. Yet the subject is too ramified to allow a short summary of either; I must leave it to the reader to find out for himself.

A recent paper by HAVAS [12] deals with ITRs between Newton's and Einstein's theory of gravitation and must have been written at about the same time as my article for *Synthese* which forms Part II of the present paper. Though different in approach, the two papers agree in their conclusions.

A feature of particular interest in Havas' paper is this. In search for a mathematical formulation of Newton's theory that could be considered a *formal anticipation* (in the sense explained) of Einstein's theory the paper treats Newtonian space-time, which is the direct product $T \times E_3$, as a nonmetrical affine 4-space with singular metric $g_{\mu v}$ so as to allow a generally covariant-4-tensor representation of Newtonian theory. Thus, in the search for formal anticipation, a *new* formulation of an old theory is found, in addition to the many ones already known before.

NOTES

* Reprinted from 'Intertheory Relations, Part I: Towards a General ITR Theory – Introduction to a New Field of Study', in *Induction, Physics, and Ethics* (ed. by P. Weingartner and G. Zecha), Dordrecht 1970; Part II is reprinted here as Chapter XV, Part III as Chapter XXII.

The following abbreviations are used:

ITR = Intertheory relation(s)
o-PM = classical point mechanics
c-PM = c-mechanics (relativistic point mechanics)
h-PM = quantum mechanics (of mass points)
QM = quantum mechanics.

[1] Tisza, L., 'The Conceptual Structure of Physics', *Reviews of Modern Physics* **35** (1963) 151–85.
[2] The subject matter of *Mathematical Logic*, a subject taught at some universities, is the logic of mathematics rather than *mathematical logic* which is logic treated in a mathematical way or, equivalently, mathematics admitting logical models$_2$.
[3] Strauss, M., 'Mathematics as Logical Syntax – A Method to Formalize the Language of a Physical Theory', *Journal of Unified Science (Erkenntnis)* **7** (1938); [Chapter VI of this volume].
[4] Strauss, M., 'Zur Begründung der statistischen Transformationstheorie der Quantenphysik', *Sitzungsberichte der Berliner Akademie der Wissenschaften, Physikalisch-mathematische Klasse* **27** (1936) 382–98; [Chapter XVII of this volume]; 'Grundlagen der modernen Physik', in *Mikrokosmos-Makrokosmos* Vol. II (ed. by H. Ley and R. Loether), Berlin 1967; [Chapter XX of this volume].
[5] With $c \to \infty$ the Minkowski invariant $(\Delta S)^2 = c^2(\Delta T)^2 - (\Delta L)^2$ does not split up into the two invariants of Newtonian space-time but becomes infinite.
[6] For a formulation of c-kinematics with redundant parameters cf. M. Strauss, 'On a Generalized Lorentz Transformation', *Annalen der Physik* **16** (1965) 105–13.
[7] According to present terminology the two 'systems' are models$_1$ rather than theories.
[8] Landé, A., *New Foundations of Quantum Mechanics*, Cambridge 1965.
[9] Cf. ref. 4.
[10] Strauss, M., 'The Lorentz Group: Axiomatics – Generalizations – Alternatives', *Wissenschaftliche Zeitschrift der Friedrich-Schiller-Universität Jena, Mathematisch- Naturwissenschaftliche Reihe* **15** (1966) 109–18 [Chapter XIV of this volume], and ref. 4, Part III.
[11] Cf., e.g., H. Grad, 'Levels of Description in Statistical Mechanics and Thermodynamics', and Edwin T. Jaynes, 'Foundations of Probability Theory and Statistical Mechanics', in *Delaware Seminar in the Foundations of Physics* (ed. by M. Bunge), Berlin-Heidelberg-New York 1967.
[12] Havas, P., 'Foundation Problems in General Relativity', in *Delaware Seminar in the Foundations of Physics* (ed. by M. Bunge), Berlin-Heidelberg-New York 1967.

PART C

FOUNDATIONAL STUDIES – SPECIAL

CHAPTER XIII

ON THE LOGIC OF 'INERTIAL FRAME' AND 'MASS'*

I shall talk about two basic concepts of physics which – though of very different nature – are not entirely disconnected even though the connection is different in different theories. The two concepts are 'inertial frame' and 'mass'. The theories we are going to consider are Newtonian mechanics, c-mechanics (mechanics of Special Relativity), Einstein's theory of gravitation, and the theory of fundamental particles (which is still in the making); h-mechanics (quantum mechanics) and h-c-theory (quantum field theory) will not be considered until the end. c-theory will play a particularly important role: all too often Newtonian mechanics and Einstein's theory of gravitation are directly contrasted with one another; this obscures the fact that c-mechanics is intermediate between the two, not only historically but also mathematically and logically.

By 'logic' I mean [the union of] syntax and semantics, so that both the formal and the material aspects of concepts are covered. Most investigations concerning the logic of physics neglect one or the other aspect. Mathematical physicists usually pay little attention to semantics while experimental physicists usually concentrate on coordinating or 'operational' definitions and overlook that the meaning of theoretical concepts is to conform to the mathematical syntax which may include implicit definitions, symmetry properties, invariance groups, etc.

I. THE CONCEPT 'INERTIAL FRAME'

Let us start with the question: how is the concept 'inertial frame' to be defined in Newtonian mechanics?

As well-known, a *kinematic* definition is not possible within Newtonian mechanics; only Newton's *equations of motion* distinguish the class of inertial frames. I call this the *discrepancy between kinematics and dynamics in Newtonian mechanics*. This discrepancy can be considered as the logical substance [or rather the proper target] of Huygens', Leibniz', and Mach's criticism of Newtonian mechanics. It may also be one of the reasons why Newton introduced the concept of 'absolute space' [im-plying the existence of a *single* preferential frame] although this concept plays no role whatsoever in his mechanics.

How does the discrepancy come about? First answer: kinematics – unless coinciding with dynamics as in Einstein's theory of gravitational motion – is an abstraction from dynamics, and in the process of ab-straction something gets lost; the thing that gets lost is here the distinction between inertial and non-inertial frames. This of course is the trivial part of the answer. We must enquire *why* this distinction gets lost. The answer: because the invariance group of Newton's dynamics is the Galileo group [– a small subgroup of the invariance group of Newtonian kinematics]. The explanation is as follows. Call the class of frames connected by the transformations of a kinematic group *kinematic equivalence class*. Let $\{\sum_1^i\}$ be frames of a Galilean equivalence class. Construct the class $\{\sum_2^i\}$ satisfying the condition that $\sum_2^i$ moves with a constant acceleration with respect to $\sum_1^i$, the acceleration having the same value and direction for all values of i. You find that $\{\sum_2^i\}$ is again a Galilean equivalence class. As the accele-ration vector has three independent arbitrary components, you can con-struct a 3-fold infinity of Galilean equivalence classes to a given class of this kind. [One of these classes is then singled out as the class of inertial frames by Newtonian dynamics; but this class cannot be defined in an explicit way, as shown below.]

Logically, there exist two possible ways to get rid of the discrepancy between kinematics and dynamics. The first way would be to *change dynamics* in such a way that [[...]] [all frames *kinematically* equivalent become also *dynamically* equivalent]. This in essence is the demand made by the 'kinematical relativists' [REICHENBACH] [1] Huygens and Leibniz. The second way would be to *change kinematics* in such a way that the new kinematic invariance group admits only *one* uniform motion equivalence class that could then be identified with the class of inertial frames.

Physics has gone the *second* way: the Lorentz group defines exactly one kinematic equivalence class which happens to be a uniform motion equivalence. This may be proved either geometrically [in Minkowski's space-time] or algebraically. From the algebraic point of view the es-sential point is that the Lorentz group is an *irreducible* representation of the algebraic velocity group in x-t-space, while the Galileo group is a *reducible* representation.[2]

The kinematical distinction of the inertial frames in c-theory may be formulated without mathematical terms as follows: inertial frames are precisely those frames in which the 4-dimensional 'lightgeometry' coincides with the chronogeometry of rigid rods and mechanical clocks. [[...]] This formulation can also be used as an 'operational definition' [viz., an operational criterion] of 'inertial frame'.

In Newtonian theory a finite limiting velocity c does not exist and hence the class of inertial frames can neither be distinguished kinematically nor be defined dynamically.

Indeed, the current 'definition' reads:

> 'A frame $(\sum)$ is an *inertial frame* (I) iff a *force free* (ff) mass point (P) *moves uniformly* (mu) with respect to $\sum$.'

Yet for 'force free' we only have the 'definition':

> 'A mass point P is *force free* (ff) iff it moves uniformly (mu) with respect to an *inertial frame* (I).'

Thus the definitions are circular [besides not being quite correct logically], and have to be replaced by axioms. Obviously, only the following three axioms hold [in Newtonian mechanics]:

(1) $\text{ff}(P) \wedge \text{mu}(P; \sum) \supset I(\sum)$
(2) $\text{ff}(P) \wedge I(\sum) \supset \text{mu}(P; \sum)$
(3) $I(\sum) \wedge \text{mu}(P; \sum) \supset \text{ff}(P).$

From these axioms, an *explicit* definition of 'inertial frame' ('I') *cannot* be deduced. Hence, 'inertial frame' and 'force free' are *primitive* (irreducible) concepts in Newtonian mechanics, connected by axioms and thus only incompletely defined. Nothing is changed [in this conclusion] if, following L. Lange, the predicate 'moves uniformly with respect to $\sum$' ['mu(; $\sum$)'] is replaced by 'moves in a straight line with respect to $\sum$ for three independent directions'. That *in praxis* this incomplete definition proves sufficient results solely from the fact that in almost all practical cases we are able to distinguish uniquely between *genuine forces* and *apparent forces* (forces of inertia, inertial forces). This in turn results from the fact that we know all macroscopic force laws.

There is a *second discrepancy* in Newtonian mechanics. It results from the fact that the force of gravity, in contrast to all other forces, is *universal* [i.e. acting between all forms of matter] and *not to be screened off* (unabschirmbar); this makes gravitation a 'form of existence' of matter and thus brings it into line with space and time. This of course is the very basis of Einstein's 'geometrization' of gravitation but it also implies that a geometrization of other forces is impossible.[3]

Let us now investigate the consequences within Newtonian theory of these two fundamental properties of gravitation.

For this purpose we introduce the following notions. A frame of reference is to be called a [*Newtonian* or] *potential inertial frame* iff Newton's law of motion (without inertial forces) holds in it. A potential inertial frame is to be called [a *Galilean* or] an *actual inertial frame* if it admits force free motion.

With these notions, the situation can be described as follows: Newtonian mechanics admits an infinity of *potential* inertial frames, connected by the Galileo group, but it does not admit a single *actual* inertial frame of finite extension in all *three* dimensions [except in empty universe] since the gravitational forces can compensate each other, if at all, only on a *two*-dimensional interface. [Obviously, this is a *discrepancy* between Newton's *general law of motion* which admits an infinity of global actual inertial frames and his *theory of gravitation* which does not admit any such frame.] From the logical point of view it can [also] be considered as a discrepancy between syntax and semantics since what is admitted by the mathematical formalism is excluded by the physical interpretation [which includes the two basic properties of gravitation].

As well-known, this second discrepancy of Newtonian mechanics has been removed in Einstein's theory of gravitation by admitting only *local* inertial frames [often called *local Galileo frames*]. However, it has not always been realized that this is not a new physical statement but the adaptation of the mathematical formalism to a state of affairs already existing in Newtonian mechanics [and expressed there by the discrepancy discussed above]. As far as pure gravitation is concerned the novel feature of Einstein's theory as compared to that of Newton resides in the *retardation* of gravitational action resulting from [the finite value of] the limiting velocity c. This explains why Einstein's theory, in spite of its very different mathematical and logical structure, usually yields only small deviations from Newtonian results, deviations that disappear for $c \to \infty$, such as Einstein's motion of the perihelion. This underlines the remark made in the beginning: the way from Newton's to Einstein's theory of gravitation leads through the c-theory.

To Mach, even Newton's theory appeared to have an unnessarily complicated structure as compared to experience. No wonder that he did not like Einstein's theory which he considered to be even more 'metaphysical'. It speaks for Mach's incorruptible judgement, praised by Einstein, that he was not deceived either by the name 'Relativitätstheorie' or Einstein's letters to him. For Mach, the empiricist, Einstein's theory was a new *fall of man*, and this judgement was later confirmed by the founder of 'modern' empiricism, P. W. BRIDGMAN[4].

Nobody can say whether Mach would have been reconciled to Einstein's theory if the Freiburg papers[5] had appeared during his lifetime. These papers confirm that, contrary to the expectations of Einstein, nothing is gained for Mach's request [known as the 'Mach Principle'] by the transition from Minkowski to Riemann space. Only a *change in global topology* [transition to *closed* space] makes it possible to satisfy some version of the Mach Principle.

This solution appears to me quite satisfactory, both logically and philosophically. Unfortunately, it has the drawback that we cannot hope ever to be in a position where we can decide the question of global topology of our universe with any degree of certainty that we are used to in physics. Even if the universe is closed the information that can possibly reach us will always remain an infinitesimal fraction of the whole, and the father it comes from the more 'dated' it will be. For cosmology it would have been better if the speed of light were a few million or billion times larger.

To summarize, the discrepancy between kinematics and dynamics in the question of preferential frames, characteristic of Newtonian theory, is overcome in c-theory, but the c-theoretical solution of the problem is [in philosophical spirit and physical substance] definitely *anti-Mach* and *pro-Newton*: Newton's inertial frames are now distinguished already kinematically. The second discrepancy of Newtonian theory is overcome by Einstein's chronogeometrization of gravitation without, however, implementing Machian ideas. Only a change in global topology [transition to closed universe] makes it possible to satisfy [a somewhat sophisticated version of] the [so-called] Mach Principle. Since, in the final analysis and also in Mach's view, statements on space and time are merely expressing properties of matter, the task remains of clarifying the testable content of cosmological models.

II. THE CONCEPT 'MASS'

Turning to the concept of mass, we start again with Newtonian mechanics.

Newton himself defines 'Mass' as the product of density and volume [besides explaining that mass is a measure of the quantity of matter]. In the opinion of Mach [and most of his followers as well as of his critics] this definition is empty or *circular* since density is mass per unit volume. Though the latter is correct, Newton's definition is *not* empty: it implies that 'mass' is an *additive quantity*, at least for bodies of equal density; [as densities can be directly compared] the additivity of mass becomes testable when Newton's axioms are taken into account. These axioms, as well-known, permit to distinguish between *inert mass*, implicitly defined by the general law of motion, and *gravitational mass*, implicitly defined by the gravitational force law; neither of these implicit definitions implies that mass is an *additive* quantity. Hence, Newton's definition of mass can only be dropped if the additivity of mass is introduced by an additional axiom.

Newton's splitting up of the law of gravitational motion into a *general law of motion* and the *law of gravitational force* has likewise been criticized by Mach. In Mach's opinion, gravitational experience justifies only the following *law of relative acceleration*:

$$\mathbf{b}_{12} = k\,(m_1 + m_2)\,\mathbf{r}_{12}/(r_{12})^3 \,.$$

He does not, however, investigate whether this law has the same content as the two Newtonian laws taken together, e.g., whether it is sufficient for determining the relative accelerations in the three-body problem. Apart from this, Newton's splitting signifies a *program*: his general law of motion *reduces the task of physics* to that of finding the special force laws. [[...]]

The empirical equality or proportionality of *inert mass* and *gravitational mass* means, logically speaking, that the *extensions* of the two concepts coincide [as happens frequently in physics]. Mach however regarded the two concepts of mass as the result of a metaphysical concept doubling [that should be eliminated] and thus arrived at the formula given above. Einstein, on the other hand, took the empirical equality of the two kinds of masses as starting point for his heuristic considerations that finally led to his General Theory of Relativity. Einstein is following here the method of Newton rather than the logic of Mach. Newtonian mechanics is using the language of the differential calculus which had been invented by Newton for just this purpose. For both Newton and Einstein the maxim was: the most *fundamental* properties of matter should be accounted for already by the *syntax* of the mathematical language used [i.e., the mathematical *substructure* of the theory], and not by *equations* [mathematical *superstructure*]. This is indeed the *true secret of theoretical physics* and the *key* to its proper understanding. The fundamental thing for Newton was Galileo's law of force free motion, formulated by Newton as his *Lex prima* but in fact a mere consequence of his *Second Law*, as also noticed by Mach. For Einstein, two things were fundamental: the *equality* of *inert* and *gravitational* mass and the *universality* of gravitation. Both taken together lead, through the principle of *local* equivalence [of acceleration and gravitational force] to Riemannian geometry as the adequate mathematical language for gravitational theory. And it is no coincidence that Einstein's result exhibits a marked analogy to Newton's Second Law: the same role that 'force' plays in Newton's theory is played in Einstein's theory by the matter tensor; to determine the latter is again the remaining task left to physics.

We now turn to the question: *what contribution has Einstein's theory of gravitation made towards a solution of the mass problem?* By 'solution of the mass problem' we mean *reduction* of the concept of mass to other fundamental concepts and *deduction* of the *mass ratios* of fundamental particles from theory. The answer is: no contribution, or almost no contribution.

Mass, generalized to the matter tensor, remains a *primitive* concept also in Einstein's theory; its [specification or] reduction to other concepts is left to the other parts of physics. Of course, you can define the matter tensor by the left-hand side of Einstein's equations, just as you may regard Newton's Second Law as a definition of 'force' or Maxwell's equations as a definition of 'electric current'. Clearly, in this way you turn physics into a tautology, a mathematical game without physical content.

What about the finite cosmological models satisfying [a certain version of] the [so-called] *Mach Principle*? Do they represent an advance towards the solution of the mass problem? I don't think so. The maximum obtainable in such models is a complete determination of the universal *inertial forces* by the energy-momentum complex of matter *and* gravitation. [This result does not even touch the problem of inert *mass*, and is a far cry away even from Einstein's *second* (and weaker) *version* of the *Mach Principle* demanding *complete* determination of the metrical tensor field by the matter tensor alone.]

Incidentally, it seems to me that, quite independent of Einstein's theory, the *Mach Principle* [or rather: a rational version of it] would amount, as far as the *mass problem* is concerned, merely to a sort of *cosmological gauging of mass* of such a kind that a common *factor* – generally a space-time function – in the masses of all bodies depends on the global distribution of matter in the universe, and that this factor should be zero for a single lone particle. This conception seems to be vindicated by the work of HOYLE[6], BRANS and DICKE[7], and GÜRSEY[8].

The mass *ratios* of the fundamental particles can never be explained in this way; this problem lies outside the competence of gravitational theory.

We now turn to *c*-theory.

In *c*-theory the Newtonian concept of *inert mass* splits up into 'rest mass' and 'relative mass'. The invariant rest mass m can be defined by

$$\sum_1^4 P_k P_k = - m^2 c^2 \quad \text{or} \quad \overset{\circ}{E} = mc^2,$$

[the second equation being a consequence of the first]. These equations [[...]] follow from the definition of the energy-momentum 4-vector and do *not* represent a reduction of the mass concept; rather do they say that 'rest mass' and 'rest energy' are synonyms. Of course, this linguistic rule – as all rules of physical syntax – has some physical content, which is well-known from nuclear physics. Yet this content follows only in connection with the conservation laws for energy and momentum. [The energy conserved is the *relative* energy E].

The *relative mass* is best defined[9] by the equation

$$m_{\text{rel}} = E/c^2$$

which makes it the 4th component of the momentum-energy 4-vector, apart from a c-factor. Accordingly, particles of zero rest mass have a [non-vanishing] relative mass. If the equality of inert and gravitational mass is to be maintained, the inert mass has to be identified with the *relative* mass [and not with the *rest* mass]. This of course concerns the theory of gravitation. For the proper mass problem, i.e. the mass ratios of fundamental particles, neither gravitational theory nor c-theory, taken by themselves, have any implications. The situation changes, however, if quantum theory is also taken into account.

In nonrelativistic h-mechanics the mass of a particle remains an arbitrary parameter just as in Newton's mechanics. As a novel feature I can only quote the fact that there is a superselection rule for mass, ensuring the stability of the particles and the conservation of mass. All this is connected with the [somewhat peculiar] fact that the irreducible 'physical' representations of the Galilei group in Hilbert space are not 'true' but 'projective' representations[10].

A fundamental change in the situation is brought about by the transition to h-c-theory (relativistic quantum theory of fields). To start with, this theory permits the mutual transformation of all kinds of particles, subject to well-known conservation laws. Some of the latter are new and apply to so-called internal degrees of freedom. The novel quantities conserved are above all the additive quantum numbers baryon charge, lepton charge, and hyper charge, further the isospin. If it were possible to ascribe an energy to these internal degrees of freedom the [rest] mass of these fundamental particles could be reduced to energy, at least in principle. As a matter of fact, the mass of the fundamental particles does depend on the internal quantum numbers, and formulae relating the masses of particles belonging to the same multiplett have been established theoretically. A complete determination of all mass ratios, however, is still far away. The internal quantum numbers known so far certainly do not suffice for such a complete determination. It appears almost certain that a quite novel and different feature has to be taken into account, viz., the interaction of the particles with the physical vacuum, which is automatically contained in theories of the Heisenberg type.

In view of these – partly empirical, partly theoretical – results the following *hypothesis* may be pronounced:

> The masses of the fundamental particles are uniquely determined by the internal quantum numbers and the coupling constants – apart from a common gauge factor.

One may even go one step further. For dimensional reasons the gauge factor is a reciprocal length. This length is either a new universal constant of nature, as in Heisenberg's theory, or the product of such a constant and a numerical gauge factor; the latter may depend on cosmology – in agreement with [a weak version of] the [so-called] *Mach Principle*.

Would such a [complex] solution of the mass problem contradict Mach's ideas? I don't think so. Mach was not committed to the Principle carrying his name. Let's hear what he wrote:

It would be quite possible that the isolated bodies *A*, *B*, *C*,,... play only an accidental role in the motion of body *K* and that the motion of *K* be determined by the surrounding medium. ... Even if this idea proves impracticable at present there is still the hope that we may learn more about this hypothetical medium – now known as physical vacuum – in the future ....[11]

It appears that [for a solution of the mass problem] we need both [a certain version of] the *Mach Principle* and Mach's hypothetical medium. The latter, to be sure, is to be regarded as a potential or virtual reality in the sense of quantum theory. In spite of this there is a certain analogy between Mach's idea that inertia is due to interaction with the other bodies of the universe on the one hand and Heisenberg's self interacting field or the bootstrap idea on the other hand.

If you permit me to conclude my remarks with a short characterization of Mach's role in the development of modern physics I would say this: it seems to me incorrect and misleading to present Mach as a man who paved the way for modern physics, as is sometimes done. Physics has developed along lines entirely different from those desired by Mach. Yet it is true to say that Mach's criticism has helped to prepare the soil so that new ideas could take roots. Furthermore, Mach has [in a certain sense] anticipated some important results or hypothetical ideas of modern physics, if only in the conceptual framework of his time. To be sure – and this is a point of [philosophical] interest – it was precisely his insistence on finding real causes [for the observed phenomena], besides his incorruptible sobriety, [and not his general philosophy], that qualified Mach to these rather astonishing foresights.

NOTES

* Translated from 'Zur Logik der Begriffe "Inertialsystem" und "Masse"', in *Symposium aus Anlass des 50. Todestages von Ernst Mach*, Ernst-Mach-Institut, Freiburg i. Br., 1966.
[1] [H. Reichenbach, 'Die Bewegungslehre bei Newton, Leibniz und Huygens', *Kantstudien* 29 (1924) 416.]
[2] Cf. M. Strauss, 'The Lorentz Group: Axiomatics – Generalizations – Alternatives', *Wiss. Z. Friedrich-Schiller-Univ. Jena*, Math.-Nat. R. 15 (1966) 109–18 [Chapter XIV of this volume].
[3] Cf. H. Reichenbach, *Philosophie der Raum-Zeit-Lehre*, Berlin – Leipzig 1928.
[4] [P. W. Bridgman, *The Logic of Modern Physics*. New York 1927; 'Einstein's Theories and the Operational Point of View' in *Albert Einstein: Philosopher-Scientist* (ed. by P. A. Schilpp), Evanston 1949.]
[5] H. Hönl and H. Dehnen, 'Über Machsche und anti-Machsche Lösungen der Feldgleichungen der Gravitation', *Ann. Physik* 13 (1964) 201–15. [Part I] and *Ann. Phys.* 14 (1964) 271–95 [Part II]. [Cf. also H. Hönl, 'Das Machsche Prinzip und seine Beziehung zur Gravitations-theorie Einsteins' in *Einstein-Symposium* (ed. by H.-J. Treder), Berlin 1966.]
[6] F. Hoyle and J. V. Narlikar, *Proc. Roy. Soc.* A 270 (1962) 334.
[7] C. Brans and R. H. Dicke, *Phys. Rev.* 124 (1961) 925.
[8] F. Gürsey, 'Reformulation of General Relativity in Accordance with Mach's Principle', *Ann. of Phys.* 24 (1963) 211–42.
[9] [For particles of non-zero rest mass this definition agrees with the usual one, viz.

$$m_{\text{rel}} = m/[1 - v^2/c^2]^{1/2}.$$

The latter definition, however, says nothing about the relative mass of particles with zero rest mass since the limit ($m \to 0$, $v \to c$) gives the indeterminate value 0/0.]
[10] Cf. J.-M. Levy-Leblond, *J. Math. Phys.* 4 (1963) 776 and the literature quoted there.
[11] E. Mach, *Die Mechanik in ihrer Entwicklung historisch-kritisch dargestellt*, 7th. Edition, Leipzig 1912, Chapter 2, Section 6.

THE LORENTZ GROUP: AXIOMATICS – GENERALIZATIONS – ALTERNATIVES*

INTRODUCTION

The 60th anniversary of Einstein's Special Theory seems to be an appropriate occasion to ask ourselves the question: how do we look upon this theory today, in particular, do we still believe in

(i)   the *truth* of the two principles used by Einstein to establish the Lorentz group *(LG)*, and

(ii) the *necessity* of using them for this purpose.

As for the first question, hardly anyone doubts the truth of the two principles, except that they must be generalized to take account of gravitation.

As for the second question, the answer is quite different. True, the textbooks, and even the monographs, go on repeating the old story. Yet axiomatics has long found an entirely different answer: of the two principles the second is not required at all, and of the first only a very weak implication is required that amounts to little more than a definition of constant speed. On the other hand, the tacit assumption of a Euclidean space proves quite essential for establishing the LG.

There is another point that requires reconsideration: the importance attached by Einstein to his analysis of the concept of simultaneity. On this, Einstein writes in his popular exposition of the theory [1] – I quote by memory: – "do not proceed until you have understood this point".

No less emphatic is H. WEYL [2]:

In der Befreiung von diesem Dogma (der objektiven Bedeutung der Gleichzeitigkeit) liegt die große erkenntnistheoretische Tat Einsteins, die seinen Namen neben den des Kopernikus rückt.

Now what does emerge from Einstein's analysis is the conclusion that 'velocity' and 'simultaneity' are interrelated concepts. From this it does *not* follow that simultaneity is a three-thing relation, viz., a relation between two events and a frame of reference (as it actually is), nor does it follow that the relation of simultaneity is a subject for arbitrary definition, as implied by Weyl and other authors. As a matter of fact, if we replace the Einstein definition of simultaneity by a different (non-equivalent) one we obtain a different physical theory as will be shown below. Moreover, even the mere interrelatedness of the two concepts follows in the Einstein analysis only if the traditional operational definition of velocity in terms of measurements of length and time is adopted. Yet we are free to use any other operational definition such as the definition by means of the Doppler effect, or no operational definition at all but an implicit definition by a set of axioms.

Thus, it will be seen from these two examples that the Special Theory is no less in need of being freed from mistaken conceptions than was the case with the General Theory. For the latter this task has been accomplished by Fock. To achieve a corresponding aim for the Special Theory is the main concern of this lecture. As the mathematical groundwork is scattered over many years this lecture will largely be a synthetic report on, or a critical examination of, the work done by various authors. Some of them are following rather different lines of thought but all will be seen to contribute to the final picture, sometimes quite unintentionally and sometimes even *against* their own intentions.

It will be best to divide the lecture into three main parts:

I   Axiomatics of the LG
II   Attempts at generalizing the LG
III   Attempts to offer alternatives.

Only in the last two parts shall I quote recent work not yet published.

I. AXIOMATICS OF 'LG'

1. *The Light-Geometrical Line*

The two postulates used by Einstein are of very different character: one a universal statement, the other concerned with light only. Thus, it is not surprising that two lines of axiomatics have developed giving preference to the one or the other of the two principles. We may call them the *algebraic or group theoretical* line and the *light-geometrical line*, respectively. The latter one aims at building up the theory exclusively with the help of light signals. Its main exponents are REICHENBACH [3],

MILNE [4] and PAGE [5]. Their work is quite interesting and in part very ingenious, but in the final analysis it merely serves to show the limitation of this approach. Thus, Reichenbach cannot dispense with rigid rods and Milne, who does, does not obtain the LG but a whole set of groups.

The reason for the ultimate failure of this line is rather obvious: it is the fact, first discovered by CUNNINGHAM [6], that the Maxwell equations, and hence the wave equations for light, in vacuo are invariant under the full (15 parameter) *conformal group* and not only under the (10 parameter) LG. Thus, in the sense of Klein's Erlanger Program, the 'space' of light-kinematics is a conformal space, and not the pseudo-Euclidean space of Special Relativity.

This conclusion has recently been confirmed from a somewhat different point of view by CASHMORE [7]. This author has proved that the most general *integrable* point transformations between moving observers that preserve a universally constant speed are of the form

$$\frac{\bar{x}^\mu - \bar{a}^\mu}{\sum\limits_\nu (\bar{x}^\nu - \bar{a}^\nu)^2} = AB^\mu_\lambda \frac{x^\lambda - a^\lambda}{\sum\limits_\nu (x^\nu - a^\nu)^2} + C^\mu \quad \text{with} \quad B^\mu_\lambda B^\mu_\sigma = \delta_{\lambda\sigma}. \tag{1}$$

This transformation contains $1 + 3 \times 4 + (16 - 10) = 19$ parameters and corresponds to the case where the line element of either observer is of the form

$$\mathrm{d}s^2 = K^2 \frac{\mathrm{d}x^\mu \, \mathrm{d}x^\mu}{\left[\sum\limits_\nu (x^\nu - a^\nu)^2\right]^2}. \tag{2}$$

Now if one starts with the usual form of the Maxwell equations or the wave equation one tacitly assumes that the line element for one observer is of the form

$$\mathrm{d}s^2 = K^2 \, \mathrm{d}x^\mu \, \mathrm{d}x^\mu \tag{3}$$

so that four parameters disappear. This reduces the number of parameters to 15 and corresponds to the case investigated by Cunningham. Furthermore, it can be shown directly that in (1) 4 of the 8 parameters $a^\mu, \bar{a}^\mu$ are redundant in the sense that we get the same manifold of transformations if either the a^μ or the $\bar{a}^\mu$ are treated as constants instead of as group parameters. The transformations (1) are then but a special way of writing the conformal transformations. Thus, we always get the conformal group if we use light only in constructing a kinematics.

Now we may agree that for a proper comparison with the Lorentz group we should impose the condition that the transformation be one between observers in *uniform relative motion*. If we use the customary definition of velocity, this condition is equivalent to demanding that the transformations be *linear*. Then the 8 parameters $\bar{a}^\nu$, a^ν are all zero and we are left with exactly *one additional parameter* as compared to the LT.

Let me summarize: *Any kinematics based solely on* Einstein's Second *Postulate is bound to lead to a* $(n+5)$ *or a* $(n+1)-parameter$ *(conformal) group or subgroup if the corresponding LG or subgroup has n parameters.* The reduction of $(n+5)$ to $(n+1)$ is only possible if we introduce definitions foreign to the spirit of light-geometry, i.e., if we borrow from ordinary physics. Even then we can never obtain the Lorentz group without further postulates.

2. *The Algebraic Line: (1) Transformation Theoretical Approach. (Space and Time as Primary Concepts)*

Though I cannot speak for all authors [8]–[11] following this line I believe that all of them started with a very simple observation: in the LG the constant c plays the role of a limiting velocity; whether this limit exists physically or not (like the absolute zero of temperature) is immaterial for the mathematical structure of the theory. Hence the LG *should be derivable without invoking any properties of light*.

The same conclusion is reached if you start from relativistic dynamics. It teaches you that if there exist particles of zero rest mass these particles would move with the limiting velocity; but it does not say that such particles exist in nature. In short: – the theory *admits* but does *not demand* the existence of light. In a world without light you would determine the magnitude of c from any of the so-called relativistic effects, just as Planck's constant h has been determined from quantum effects and not by measuring any quantity having the magnitude h.

Now if the Second Postulate is not required we should be able to derive the LG either from the First Postulate alone or from a combination of the First Postulate and some further postulates essentially weaker than the Second. The precise position is this: – leaving aside purely formal generalizations not affecting the physical content of the theory the First Postulate *does* suffice to derive the LG, except for one thing: the *sign* of an invariant quantity remains *undetermined*: $I = \pm c^2$. If you choose the 'wrong' sign, c is a certain critical velocity in a closed space [12].

I shall now give you an outline of the proof without mathematical details which are rather trivial, but emphasizing the physico-logical points. The first and most important step is to realize that the First Principle, known as the Principle of Relativity or the Principle of Kinematic Equivalence, implies this: the transformations between equivalent frames form a *group* with the *relative velocity as group parameter*; it also implies that all fundamental laws are covariant (form invariant) under this group, but for kinematics only the first (much weaker) implication is required.

Now write the transformation between any two frames

$$\overset{i}{\sum} = \{t^{(i)}, x^{(i)}, y^{(i)}, z^{(i)}\}$$

and

$$\overset{k}{\sum} = \{t^{(k)}, x^{(k)}, y^{(k)}, z^{(k)}\}$$

in the symbolic form

$$\overset{i}{\sum} = L^{ik} \overset{k}{\sum} \quad \text{(no summation!)} \tag{4}$$

with some operator

$$L^{ik} = L(v_k^i). \tag{5}$$

v_k^i is the velocity between $\sum^i$ and $\sum^k$ as measured in either $\sum^i$ or $\sum^k$; for the sake of being definite, let us say: as measured in $\sum^i$. Then the equivalence of the frames is expressed by the three group properties

$$L^{ii} = L(v_i^i) = L(o) = I \text{ (identity operator)} \tag{6}$$

(which is trivial),

$$L^{ki} = (L^{ik})^{-1} \quad \text{i.e.} \quad L(v_i^k) = L^{-1}(v_k^i) \tag{7}$$

and

$$L^{ik}L^{kl} = L^{il} \quad \text{i.e.} \quad L(v_k^i) L(v_l^k) = L(v_l^i). \tag{8}$$

The last equation implies that v_l^i is some function of v_k^i and v_l^k, which we call the *physical sum* of v_k^i and v_l^k:

$$v_l^i = v_k^i \dot{+} v_l^k = v_m^i \dot{+} v_l^m. \tag{9}$$

So far I have not introduced any restriction on the frames, except that their relative velocity be constant. From now on I shall impose the usual restrictions, viz., I shall assume that the $x^{(i)}$, $y^{(i)}$, $z^{(i)}$ etc. are Cartesian coordinates, that the $x^{(i)}$-, $y^{(i)}$-, $z^{(i)}$-axes are prallel (and not anti-parallel) to the $x^{(k)}$-, $y^{(k)}$-, $z^{(k)}$-axes, respectively, that for $t^{(i)} = t^{(k)} = 0$ the origins of the two frames coincide, and, finally, that the relative velocity has the direction of the x-axes. In this way we get rid of the 4 translations, the 3 spatial rotations, and 2 of the 3 velocity parameters. Then we should expect that our v is the *only* group parameter. Before I come back to this question let me point out a conclusion that is independent of how this question is decided.

If you reverse either the t-axes or the x-axes you change the signs of all velocities. Hence Equation (9) implies

$$-(v_1 \dot{+} v_2) = (-v_1) \dot{+} (-v_2). \tag{10}$$

This gives you the first functional equation for the unknown function $\dot{+}$. Now the next step is to realize that you need not consider the case where you have other parameters besides v: any such additional parameters would not be related to anything existing in nature but would merely reflect some clumsy convention regarding the comparison of standards in different frames. Thus, you may say that *by definition* (and not by any further physical axiom) v *is the only group parameter left*. Hence, by a well-known theorem of group theory, you conclude that the transformations (4) *commute*:

$$L(v_1)\, L(v_2) = L(v_2)\, L(v_1) \tag{11}$$

or, equivalently,

$$v_1 \dot{+} v_2 = v_2 \dot{+} v_1. \tag{12}$$

The next step is to introduce a new postulate viz.,

P 1.
$$v_b^a = - v_a^b$$
(13)

or equivalently,

$$L(v)^{-1} = L(-v)$$.
(14)

You would like to deduce this from the previous results, but you can't. You would not even be allowed to introduce P 1 if the transformations were known not to commute, because the commutative law is implied by (13) and (10). Indeed, (10) means

$$- v_l^i \equiv - (v_k^i \overset{.}{+} v_l^k) = (- v_k^i) \overset{.}{+} (- v_l^k).$$
(10a)

Hence, (13) implies that

$$v_i^l = v_i^k \overset{.}{+} v_k^l$$

whereas by definition

$$v_i^l = v_k^l \overset{.}{+} v_i^k .$$

Thus, P 1 is not only a new postulate but it is so strong that it implies the commutative law. On the other hand, it is a *pure convention from the physical point of view*: you can always adjust the time scale in $\sum^a$ or $\sum^b$ such that P 1 is satisfied.

Surely, this is a rather strange state of affairs which calls for elucidation. [The elucidation will emerge from the discussion in the next Section.] Incidentally, Einstein also uses P 1, but he takes this relation for granted, i.e., logically necessary, which it is not.

All previous results you can deduce without presupposing that the transformations are linear. But of course they must be linear if the velocities are to be constant.

Thus, L^{ik} is a matrix

$$L^{ik} = [l_{\mu\nu}^{ik}] = [l_{\mu\nu}(v_k^i)]$$
(15)

and all you have to do is to determine the 4×4 functions $l_{\mu\nu}(v)$. If you use

Cartesian coordinates you will realize that

$$[l_{\mu\nu}(v)] = \begin{bmatrix} l_{00}(v) & l_{01}(v) & 0 & 0 \\ l_{10}(v) & l_{11}(v) & 0 & 0 \\ 0 & 0 & 1 & 0 \\ 0 & 0 & 0 & 1 \end{bmatrix}. \tag{16}$$

Now the remaining four functions are uniquely determined by (11) and (14). In particular you will deduce from (11) that the ratio of two of the functions is an invariant, i.e., independent of the group parameter:

$$\frac{l_{10}(v_1)}{l_{01}(v_1)} = \frac{l_{10}(v_2)}{l_{01}(v_2)} = \text{invariant} \equiv \sigma c^2 \qquad (\sigma = \pm 1). \tag{17}$$

Thus you get an invariant speed as a consequence of the First Principle. Naturally you cannot exclude the case $c^2 = \infty$, but this is a degenerated case rather than an alternative. Thus, in addition to the First Principle (P 0) and (P 1) you require only two very weak postulates:

P 2 $\qquad c^{-2} \neq 0$ $\hfill (18)$

P 3 $\qquad \sigma \neq -1$, i.e., $\sigma = 1$. $\hfill (19)$

If you choose the 'wrong' sign $(\sigma = -1)$ you get for the composition of collinear velocities

$$\beta_l^i = \frac{\beta_k^i + \beta_l^k}{1 - \beta_k^i \beta_l^k} \quad (\beta = v/c). \tag{20}$$

This looks almost absurd, not because the velocities v can now take any value between $-\infty$ and $+\infty$, but because β_l^i jumps from $+\infty$ to $-\infty$ when $\beta_k^i \beta_l^k$ goes continuously from $1 - |\varepsilon|$ to $1 + |\varepsilon|$ $(\varepsilon \to 0)$. From this you would conclude that the velocity space (and hence the physical space) is closed at infinity.

3. *The Algebraic Line: (2) Relation Theoretical Approach (Velocity as Primitive Concept)*

Now let me present a somewhat different approach which I discovered some years ago [12] and which I consider more adequate than the previous one. Historically, the motive behind this new approach was the question, suggested by elementary particle physics, whether it would not be possible to generalize the LG by admitting a second universal constant of dimension of length or time to enter the equations. What I found was not a generalization but a new proof of LG.

Let us start with a critical examination of the transformation theoretical approach just presented. If you use this approach you regard space and time as primary (primitive, undefined) concepts and velocity as a secondary (derived, defined) concept.

This, however, leads to a serious defect in the definition of velocity: you cannot derive any connection between the two velocities

$$\mathbf{v}^i_{k\,\text{Def.}} = \left(\frac{\mathrm{d}\mathbf{r}^i}{\mathrm{d}t^{(i)}}\right)_{\mathrm{d}\mathbf{r}^k = 0} \tag{21a}$$

$$\mathbf{v}^k_{i\,\text{Def.}} = \left(\frac{\mathrm{d}\mathbf{r}^k}{\mathrm{d}t^{(k)}}\right)_{\mathrm{d}\mathbf{r}^i = 0} \tag{21b}$$

although both refer to the same physical situation.

It is precisely this defect which made it necessary to introduce the postulate P 1.

Now the reason for this defect is quite obvious: although velocity is a symmetrical relation the definitions (21) are non-symmetrical. They are non-symmetrical because they refer to length and time, to be measured in one or the other of the two frames, respectively. There is further an even more serious defect. Consider the case of two point particles. Then you cannot even apply the definitions (21) unless you first construct space-time frames in which the particles are at rest. But the state of relative motion of the two particles exists, whether you construct the frames or not. Thus, the space-time frames should be quite irrelevant to a properly defined notion of constant velocity.

You may wish to compare this analysis with Einstein's analysis of the concept of velocity. Einstein takes the definitions (21) for granted; he doesn't bother about the two-fold mathematical representation of a single physical relation, but he notices that t^i is not the local but the extended or frame time in $\sum^i$, which involves a definition of simultaneity in $\sum^i$. Hence he concludes that velocity (in $\sum^i$) and simultaneity (in the same $\sum^i$) are interrelated concepts – a conclusion that is quite correct but leaves no trace in the final theory (we shall see a trace of it in a different theory to be discussed below). On the other hand, the analysis given above concerns the irritating fact that we first represent a *single* physical relation (relative velocity) by *two* mathematical quantities, *independent of one another by definition*, and then correct for this logical blunder by introducing a separate postulate (P 1).

The conclusion to be drawn is this: give up trying to define velocity in terms of length and time and, instead, *take the relation of constant velocity as a primitive notion to be characterized ('defined implicitly') by a set of axioms.* This, by the way, is the normal procedure in axiomatics.

What about the Equations (21) connecting space-time notions with velocity? Well, you do need them for constructing a transformation theory in space-time, but you use them only for defining metrical coordinates. To be more specific: if you consider the metrical relation 'of equal length in $\sum^i$ and $\sum^{k'}$ as already operationally defined (e.g. by transport of rods or by means of wavelengths), then Equations (21) would define the extended or frame times $t^{(i)}$ and $t^{(k)}$. Note that the Einstein question of how to define simultaneity in $\sum^i$ does not even arise here: it is implicitly answered by the definition of frame (extended) time $t^{(i)}$. The postulates you need for characterizing the relation of constant velocity are

$$\text{A 1} \qquad v_{ik} = v_{ki} \qquad \text{(symmetry)} \tag{22}$$

$$\text{A 2} \qquad v_{ik} \dotplus v_{kl} = v_{il} \quad \text{(transitivity)} \tag{23}$$

$$\text{A 3} \qquad -(v_1 \dotplus v_2) = (-v_1) \dotplus (-v_2). \tag{24}$$

Note that A 1 and A 2 now *imply*

$$\boxed{v_1 \dotplus v_2 = v_2 \dotplus v_1} \tag{25}$$

Here, v_{ik} is just *the velocity between P_i and P_k*.

You now introduce the notion *velocity between P_i and P_k, judged by P_i,* by the definition

$$v_k^i = \sigma_i v_{ik},$$

with

$$\sigma_i^2 = 1, \quad \sigma_i / \sigma_k = -1. \tag{26}$$

You then obtain from A 1:

$$\boxed{v_k^i = -v_i^k} \tag{27}$$

from A 2, A 3:

$$\boxed{v_k^i \dot{+} v_l^k = v_l^i}.$$ (28)

Thus you have *derived* not only P 1 (27) but also all other relations for v_k^i.

I shall not repeat here how the LG (with $\pm c^2$) follows from these postulates and the definitions (26), (21); the mathematics required is somewhat more involved: instead of doing simple algebraic calculations you have to solve functional equations.

It may be, however, that from a sufficiently abstract point of view an equivalence and a group are the same thing. It would even suffice to know that a one-parametric group is completely determined by the algebraic properties of the composition law, i.e. of the function $\dot{+}$; then the LG would follow directly from A 1−A 3 and the definition (26).

II. GENERALIZATIONS

The generalizations I am going to discuss are of a purely formal character and result from the freedom in definition. This anyhow is the opinion of their respective authors [13], [15] – whether this opinion is entirely correct remains to be seen.

4. *First Generalization: 'Arbitrary Units in Different Frames'* [13]

Let L be the matrix of the ordinary LT given by (16). If you define

$$\hat{L}^{ik} \equiv \underset{\text{Def.}}{\hat{L}(\alpha_k^i, \beta_k^i)} = \alpha_k^i L(\beta_k^i) \equiv \alpha_k^i L^{ik}$$ (29)

you find that the transformations $\hat{L}^{ik}$ form a 2-parameter group with the composition laws

$$\beta_k^i = -\beta_i^k, \quad \beta_k^i = \beta_l^i \dot{+} \beta_k^l \quad \text{(as before)}$$ (30)

and

$$\alpha_k^i = -\alpha_i^k, \quad \alpha_k^i = \alpha_k^i \alpha_k^l.$$ (31)

There is a simple interpretation of this: you obtain (29) if you substitute

$$x_\mu^{(i)} = \alpha_i \varsigma_\mu^{(i)} \quad (x_0^{(i)} \equiv ct^{(i)})$$ (32)

in the ordinary LT. The laws (31) are then identically fulfilled because you have

$$\alpha_k^i = \alpha_i / \alpha_k \tag{33}$$

as a consequence of (32).

Now you may interpret (32) physically as a splitting up of the dimensional variables $x_\mu^{(i)}$ into a dimensional unit α_i and the numerical variables $\xi_\mu^{(i)}$. Then the α_k^i are nothing but the familiar conversion factors between the different units of length used in $\sum^i$ and $\sum^k$, respectively.

Quite obviously, the generalization obtained is of a purely formal character as long as we stick to the physical interpretation just given.

You can obtain a further generalization by allowing different units to be used in different frames for time or velocity as well. You then obtain a 3-parameter group instead of a 1-parameter one, or a $(n+2)$-parameter group instead of the corresponding n-parameter Lorentz group or subgroup. All this is quite trivial.

There exist however other physical interpretations such that the resulting theories are no longer physically equivalent to Special Relativity. These I have discussed in a forthcoming paper [14] which may be consulted by those interested in the problem.

5. Second Generalization: 'Anisotropic Propagation of Light' [15]

This generalization is based on the idea that you may replace the Einstein definition of simultaneity by any other one consistent with the fundamental principle of light propagation. It is thus based on Einstein's analysis of the interrelation between simultaneity and velocity.

For this reason the results are of special interest, regarding our previous discussion.

The fundamental principle of light propagation says that the local time interval between departure and return of a light signal taking a closed path is independent of the orientation of the path polygon in space, i.e.,

$$\sum \frac{|\mathbf{r}_i|}{c_i} = \sum \frac{|\mathbf{r}_i'|}{c_i'} = c^{-1} \sum |\mathbf{r}_i| . \tag{34}$$

The last equation defines an invariant speed c, viz., invariant under spatial rotations and translations. But it is assumed that c has the same value in all frames of a uniform motion equivalence. From (34) it follows that the speed of light in any direction $\mathbf{r}$ is limited by the unequality

$$c/2 < c_\mathbf{r} < \infty. \tag{35}$$

If you put $c_\mathbf{r} = c$ you get isotropic propagation of light and the Einstein definition of simultaneity. But formally you can carry through the theory with any other definition implying anisotropic propagation of light. The procedure is as follows.

You first introduce as measures of anisotropy the expressions

$$\begin{aligned}
X &= 1 - c/c_x = X_1 \\
Y &= 1 - c/c_y = X_2 \\
Z &= 1 - c/c_z = X_3 \\
R &= 1 - c/c_\mathbf{r}, \quad \mathbf{r} = (x, y, z).
\end{aligned} \tag{36}$$

These expressions are such that

$$-1 < X_\mu < 1, \; -1 < R < 1 \tag{37}$$

as a consequence of (35).

With the help of (34) you can prove that

$$rR = xX + yY + zZ \equiv x^\rho X_\rho. \tag{38}$$

Next you establish the relation between the two extended times in point $\mathbf{r} = (x, y, z)$ calculated for anisotropic and isotropic propagation of light, respectively. (In doing this you of course use the Einstein procedure of defining simultaneity by light signals, only you do not presuppose isotropic propagation of light).

The result is

$$ct_{(\mathbf{r})} = c\tilde{t}_{(\mathbf{r})} + X_\rho x^\rho, \quad r = \{x^\rho\} \tag{39}$$

or, written for an arbitrary frame $\sum^i$,

$$ct^{(i)} = c\tilde{t}^{(i)} + X_\rho^i x^{(i)\rho} \tag{40}$$

where the suffix has been dropped as no longer required. If you supplement (40) by

$$x^{(i)\rho} = \tilde{x}^{(i)\rho}, \tag{41}$$

substituting (40), (41) into the LT gives a new group of transformations
with the matrices

$$\tilde{L}^{ik} = A^i L^{ik} (A^k)^{-1},$$
(42)

$$A^i = \begin{pmatrix} 1 & -X & -Y & -Z \\ 0 & 1 & 0 & 0 \\ 0 & 0 & 1 & 0 \\ 0 & 0 & 0 & 1 \end{pmatrix}.$$
(43)

Thus, you obtain a 7-parameter transformation group instead of a
1-parameter one:

$$\tilde{L}^{ik} = \tilde{L}(\beta^i_k, X^i_\rho, X^k_\rho) \quad (\rho = 1, 2, 3).$$
(44)

However, in contrast to the previous generalization, the new parameters
X^i_ρ, X^k_ρ are not subjected to any composition law by the requirement that
the transformations (44) form a group; hence β remains the only group
parameter proper. β^i_k has not, however, the physical meaning of the
relative velocity if the latter is defined in the usual way, i.e., in terms of
$\tilde{x} = x$ and t. With

$$\tilde{\beta}^i_k \equiv \left(\frac{d\tilde{x}^{(i)}}{d\tilde{t}^{(i)}}\right)_{d\tilde{x}^{(k)}=0}$$
(45)

we get

$$\tilde{\beta}^i_k = \beta^i_k [1 - \beta^i_k X^i]^{-1}$$
(46)

and hence

$$\tilde{\beta}^k_i = -\tilde{\beta}^i_k [1 + \tilde{\beta}^i_k (X^i + X^k)]^{-1} \neq -\tilde{\beta}^i_k.$$
(47)

Now let us discuss the question whether the generalization given by (42)
is of a purely formal character (as its author claims it to be) or not.

First, it must be admitted that the general principle of light propagation stated above is all we can prove with purely optical means, or to be
more precise: if clocks are used merely to define local time, and rods
merely to define the comparison of length in one and the same frame. If
we impose this restriction on ourselves we are indeed free to define
simultaneity such that the propagation of light becomes anisotropic.
But this has very far-reaching consequences. It can be shown, for instance, that the number of periods of a clock within a giving interval of

its world line is not an invariant but depends on the frame in which the calculation is carried out, unless we choose the anisotropy parameters X_μ^i the same for all frames $\sum^i (X_\mu^i = X_\mu^k)$. Thus, the freedom of choice is already restricted by the requirement of physical consistency. This is quite contrary to the basic idea of the author.

Furthermore, even with this restriction, we would come into conflict with all laws of ordinary physics unless we give up all the usual operational definitions implied or admitted by these laws and replace them by very complicated and arbitrary ones depending on the choice of the three parameters X_ρ. Since this is contrary to the very spirit of theoretical physics we are left with the only alternative always used in such cases, viz., *we would have to reformulate all physical laws such that they become invariant under the anisotropy transformations*

$$A(X_\rho) \equiv A(X, Y, Z) = \begin{pmatrix} 1 & -X & -Y & -Z \\ 0 & 1 & 0 & 0 \\ 1 & 1 & 1 & 0 \\ 0 & 0 & 0 & 1 \end{pmatrix}. \tag{48}$$

These transformations do form a group with

$$A^{-1}(X_\rho) = A(-X_\rho) \tag{49}$$

and

$$A(X_\rho') A(X_\rho) = A(X_\rho' + X_\rho). \tag{50}$$

For dynamics and field theory it may be possible to carry out this program without loosing physical content but kinematics would definitely loose the greater part of its physical content because we *would retain only those kinematical laws that do not depend on the parameters X_ρ*. Most of *the familiar laws* of kinematics such as time dilatation, Lorentz contraction, velocity composition etc. would loose all physical content because they would be *reduced to the standard of mere definitions*.

Thus, we come to the following conclusion:

The generalized kinematics obtained by using the Einstein *procedure of defining extended frame time or simultaneity* (light signals) *but admitting anisotropic propagation of light* within the limits of the general principle of light propagation *does NOT give a kinematics physically equivalent to that of Special Relativity* (contrary to what is claimed by the author) *but a kinematics either contradicting Special Relativity (if we stick to the familiar physical definitions) or else essentially poorer* in *physical content* (if we give up the familiar physical definitions).

This confirms in a striking way the general standpoint explained above.

III. ALTERNATIVES – THE KINEMATICS IMPLIED BY PALACIOS AND GORDON [16]

To round off the picture I should now discuss alternatives to Einstein's Special Theory. Now I am not going to waste your time discussing the various attempts, appearing quasi periodically, explaining the relativistic effects, i.e. the c-effects, 'dynamically' either within the frame of absolute time (*Galilei* transformation) or on the assumption of an absolute space (privileged frame). As for the attempts in the first category, their authors just fail to realize that the *Galilei* group has never been properly established except as the degenerated case $c = \infty$ of the LG. The attempts in the second category, in so far as they make sense at all, belong to cosmology which is far too complicated a subject to be discussed here.

There is just one attempt of providing an alternative to the Special Theory that is worth discussing – not because it is a serious competitor to the Einstein theory (as their authors claim it to be) but because it illustrates a possibility not noticed before: *how symmetry may arise from asymmetry*. Incidentally, it is just this aspect of their work that seems to have escaped notice both of the authors and the critics. To stress this aspect we shall present a somewhat more general theory; the kinematics implied by *Palacios-Gordon* is just a special case of it.

From a systematic point of view the theory could have been discussed under the heading of Part I.1 (light-geometrical line in axiomatics) because it is based on the Second Principle. However, it is a theory *sui generis*; therefore I have left it to the last part of this lecture. So much by way of introduction.

It will be best to start with the theorem proved by Cashmore that the 2-parameter generalized LG with elements

$$\hat{L}^{ik} \equiv \hat{L}(\alpha_k^i, \beta_k^i) \equiv \alpha_k^i L(\beta_k^i) \tag{51}$$

is the most general group (apart from spatial rotations and translations) connecting frames in collinear uniform motion that preserves a universally constant speed c. If you now let α become a function of β:

$$\alpha = g(\beta), \quad \alpha_k^i = g(\beta_k^i) \tag{52}$$

c remains an invariant, but the resulting substitutions with the matrices

$$L_g^*(\beta) \equiv g(\beta) \, L(\beta) \tag{53}$$

do not form a group because

$$L_g^{*\,-1}(\beta) = \frac{1}{g(\beta)} \, L(-\beta) \neq L_g^*(f(\beta)). \tag{54}$$

Even if you allow the function g to vary, i.e. if you consider the substitutions

$$L^{**}(g, \beta) \equiv g(\beta) \, L(\beta) \tag{55}$$

with g as a *functional parameter* so that (54) becomes

$$L^{**\,-1}(g, \beta) = L^{**}(g^{-1}, -\beta) \tag{56}$$

you still don't get a group because the second group property is not fulfilled:

$$L^{**}(g_2, \beta_2) \, L^{**}(g_1, \beta_1) = g_2(\beta_2) \, g_1(\beta_1) \, L(\beta_1 \dot{+} \beta_2) \neq L^{**}(g, \beta)$$

viz., there exists no function g such that

$$g(\beta) \equiv g(\beta_1 \dot{+} \beta_2) = g_2(\beta_2) \, g_1(\beta_1)$$

[except in the trivial case where g_1 or g_2 is a constant or where one of the two parameters β_1, β_2 is zero].

In short, *if the independent numerical parameter α is replaced by a function g of the other parameter β, the transformations (51) cease to form a group, even if g is considered a variable parameter.*

Now consider the 2-parametric substitutions with the matrices

$$\left. \begin{aligned} \hat{G}_\sigma^{ik} \equiv \hat{G}(\beta_\sigma^i, \beta_\sigma^k) &\underset{\text{Def}}{=} L_g^*(\beta_\sigma^i) \, L_g^{*\,-1}(\beta_\sigma^k) = \\ &= [g(\beta_\sigma^i) \,|\, g(\beta_\sigma^k)] \, L(\beta_k^i) \\ \beta_k^i \equiv \beta_\sigma^i \dot{+} \beta_k^\sigma \end{aligned} \right\} \tag{57}$$

you find

$$\boxed{\hat{G}(0,0) = I} \tag{58a}$$

$$\hat{G}^{-1}(\beta_\sigma^i, \beta_\sigma^k) = \hat{G}(\beta_\sigma^k, \beta_\sigma^i) \quad \text{or} \quad \boxed{(\hat{G}_\sigma^{ik})^{-1} = \hat{G}_\sigma^{ki}} \tag{58b}$$

$$\boxed{\hat{G}_\sigma^{ik}\hat{G}_\sigma^{kl} = \hat{G}_\sigma^{il}}. \tag{58c}$$

Thus, *the substitutions $\hat{G}_\sigma^{ik}$ (for fixed σ) do form a group*. The same holds for the substitutions with the matrices

$$\check{G}_k^{\sigma\tau} \equiv \check{G}(\beta_k^\sigma, \beta_k^\tau) \underset{\text{Def.}}{=} L_g^{*-1}(\beta_k^\sigma) L_g^*(\beta_k^\tau) = [g(\beta_k^\tau) \,|\, g(\beta_k^\sigma)] L(\beta_\sigma^\tau). \tag{59}$$

Thus, starting with asymmetrical 1-parameter substitutions $L_g^*(\beta)$ that *do not* form a group you have constructed two 2-parameter groups with the elements $\hat{G}(\beta_1, \beta_2)$ and $\check{G}(\beta_1, \beta_2)$, respectively. The scheme of the construction is this:

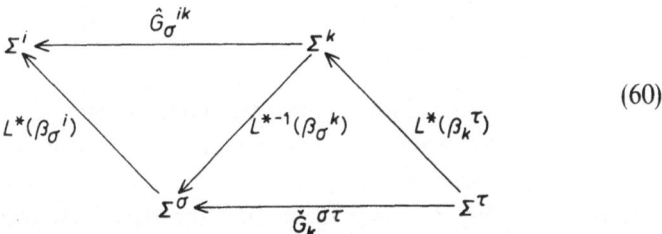

$$\tag{60}$$

Now the question arises whether you can drop the lower index in $\hat{G}_\sigma^{ik}$, i.e., whether the group properties are maintained if you use *different* indices ($\sigma \neq \rho$) in the product $\hat{G}_\sigma^{ik}\hat{G}_\rho^{kl}$.

Well, the requirement

$$\boxed{\hat{G}_\sigma^{ik}\hat{G}_\rho^{kl} = \hat{G}_\tau^{il}} \tag{61a}$$

means

$$\frac{g(\beta_\sigma^i)\, g(\beta_\rho^k)}{g(\beta_\sigma^k)\, g(\beta_\rho^l)} = \frac{g(\beta_\tau^i)}{g(\beta_\sigma^l)}. \tag{61b}$$

This introduces two new variables ($\beta_\tau^i, \beta_\tau^l$) but only one new equation.

On the other hand, you have now a second equation between the 6 variables of (61b), viz.,

$$\{\beta_\sigma^i \dotplus (-\beta_\sigma^k)\} \dotplus \{\beta_\rho^k \dotplus (-\beta_\rho^l)\} = \beta_l^i = \beta_\tau^i \dotplus (-\beta_\tau^l). \tag{62}$$

Thus, with (61), you are just left with 4 independent variables, the same number as without (61). Hence, you can drop the lower index if you pay due attention to (61b) and (62).

The next question is whether we can identify the group elements $\hat{G}^{ik}$ on the one hand with the group elements $\check{G}^{\sigma\tau}$ on the other hand. If you compare the definitions (57) and (59) you see that for identification you must have

$$\beta_k^i = \beta_\sigma^\tau \tag{63}$$

and, equating the square brackets,

$$g(\beta_\sigma^i)\, g(\beta_k^\sigma) = g(\beta_\sigma^k)\, g(\beta_k^\tau). \tag{64}$$

Since $\beta_k^\sigma = -\beta_\sigma^k$, you have 2 equations for 6 variables. In addition you have 2 more equations:

$$\beta_\sigma^i \dotplus \beta_k^\sigma = \beta_k^i$$
$$\beta_k^\tau \dotplus \beta_\sigma^k = \beta_\sigma^\tau.$$

Hence, of the 6 variables two remain independent.

You have thus proved: *the two groups $\hat{G}$ and $\check{G}$ have the same elements, viz., the same matrices.*

This, however, does *not* mean that the *frames* $\sum^\rho$ can be identified with the frames $\sum^i$. In fact, they *cannot be identified* because the transformation between a $\sum^\rho$ and a $\sum^i$ is not given by an element of $\hat{G}$ (or $\check{G}$) but by an asymmetrical 1-parameter substitution $L_g^*(\beta)$.

Let me summarize: *if you admit in kinematics an asymmetrical transformation of the type*

$$L_g^* = g(\beta)\, L(\beta), \tag{53}$$

L(β) being the ordinary LT, you imply the existence of two distinct classes of frames,

$$\hat{C} = \left\{ \overset{i}{\sum} \right\}$$

and

$$\check{C} = \left\{ \overset{\alpha}{\Sigma} \right\}$$

each class forming a separate uniform motion equivalence as expressed by the groups

$$\hat{\mathfrak{G}} = \{\hat{G}^{ik}\}$$

and

$$\check{\mathfrak{G}} = \{\hat{G}^{\rho\sigma}\}$$

respectively, (with $\hat{\mathfrak{G}} = \check{\mathfrak{G}}$ if we abstract from the frames), each frame of $\hat{C}$ being connected with any frame of $\check{C}$ by an asymmetrical transformation of the type (53).

Now we are sufficiently prepared to discuss the theory of *Palacios* and *Gordon*. This theory is just the special case

$$g(\beta) = \varkappa^{-1}(\beta) \equiv \sqrt{1 - \beta^2}.$$

By choosing this particular function you achieve that the *time dilatation* for a clock resting in a $\Sigma^\rho \in \check{C}$ disappears in a $\Sigma^k \in \hat{C}$ while the Lorentz *contraction* of a rod resting in a $\Sigma^k \in \hat{C}$ disappears in a $\Sigma^\rho \in \check{C}$.

It is clear that nothing is gained thereby in the way of getting rid of relativistic (i.e., c-) effects, which was the aim of Palacios and Gordon. On the other hand, the view expressed by Romain that this theory is physically equivalent to Einstein's is certainly mistaken: the principle of kinematic equivalence is here not fulfilled for *all* frames in uniform relative motion but only for the frames of one and the same class $\check{C}$ or $\hat{C}$. Moreover, any transformation between frames of the same (equivalence) class has *two* parameters (instead of one) referring to a frame of the other class.

Needless to say that the novel features of the theory implied by *Gordon-Palacios* – or any other theory of this type – cannot be reconciled with our physical experience.

IV. SUMMARY

(1) There exists no equivalent substitute for the Special Theory – only non-equivalent alternatives.

(2) The Second Principle is correct but not required for the axiomatic foundation of the theory.

(3) Of the First Principle only the 'weak implication' is essential for kinematics; it amounts to little more than a definition of the *uniform motion equivalence*.

(4) The empirical content of the theory lies mainly in the possibilities it excludes: unlimited speed and 'wrong sign' of c^2, the double-world implied by Gordon-Palacios, etc.

(5) The interrelatedness of *speed* and *simultaneity* in $\sum$ is irrelevant for the theory: it is based on an arbitrary and incomplete operational definition of velocity and leaves no trace in the final theory.

(6) If constant velocity is taken as primary concept characterized by a set of postulates the Lorentz group emerges as the mathematical expression of the uniform motion equivalence.

(7) There is one genuine generalization of the Special Theory: the General Theory; this theory is more general because it includes gravitation.

This has been a very critical lecture. Thus, in closing let me remind you of a classical wisdom: IM ANFANG WAR DIE TAT (Goethe, *Faust* I).

BIBLIOGRAPHY

[1] Einstein, A., *Über die Spezielle und Allgemeine Relativitätstheorie*, Braunschweig, 1917.
[2] Weyl, H., *Raum-Zeit-Materie*, 4. Aufl., Berlin, 1921, p. 156.
[3] Reichenbach, H., *Axiomatik der relativistischen Raum-Zeit-Lehre*, Braunschweig, 1922.
[4] Milne, E. A., *Kinematic Relativity*, Oxford, 1948.
[5] Page, L., and Adams, N., *Electrodynamics*, New York, 1940.
[6] Cunningham, E., *Proc. London Math. Soc.* **8** (1910), 77.
[7] Cashmore, D. C., *Proc. Phys. Soc. (London)* **81** (1963), 181.
[8] Ignatowski, W. v., (a) *Arch. Math. Phys.* **17** (1910), 1; **18** (1911), 17; (b) *Phys. Z.* **11** (1910), 972; **12** (1911), 779.
[9] Frank, Ph., and Rothe, H., (a) *Ann. Phys.* (IV) **34** (1911), 925; (b) *Phys. Z.* **13** (1912), 750.
[10] Pars, *Phil. Mag.* **42** (1921), 249.
[11] Strauss, M., 'A Simple Proof of the Lorentz Transformation', *Nature* **157** (1946), 516–17.
[12] Strauss, M., 'Grundlagen der Kinematik. I: Die Lösungen des kinematischen Transformationsproblems', *Wissenschaftliche Zeitschrift der Humboldt-Universität zu Berlin, Math.-Nat. R.* **7** (1957/58), 609–616.

[13] Romain, J. E., 'On Some Misconceptions About Relativistic Co-ordinate Transformations', *Nuovo Cimento* **30** (1963), 1254–1271.

[14] Strauss, M., 'On a Generalized Lorentz Transformation', *Annalen der Physik* (VII) **16** (1965), 105–113.

[15] Edwards, W. F., 'Special Relativity in Anisotropic Space', *Amer. J. Phys.* **31** (1963), 482–89.

[16] Palacios, P., *Rev. Acad. Cienc. Fis. Nat. Madrid* **51** (1957), 21; **165**, 245, 405; Gordon, C. N., *Proc. Phys. Soc. (London)* **80** (1962), 569.
[Cf. also Strauss, M., 'On the Voigt-Palacios-Gordon Transformation and the Kinematics Implied by It', *Il Nuovo Cimento* (X) **39** (1965), 1–9.]

NOTE

* Contribution to *Internationales Seminar über Probleme der relativistischen Physik*, organized by the *Theoretisch-Physikalisches Institut der Friedrich-Schiller Universität Jena* and held in Georgenthal, February 15–27, 1965. Reprinted from *Wissenschaftliche Zeitschrift der Friedrich-Schiller-Universität*, Mathematisch-Naturwissenschaftliche Reihe **15** (1966), 109–118.

CHAPTER XV

EINSTEIN'S THEORIES AND THE CRITICS
OF NEWTON* – INTERTHEORY RELATIONS II
An Essay in Logico-Historical Analysis

INTRODUCTION

In the history of modern physics the name of Ernst Mach has left two distinct marks, known as *Mach number* and *Mach Principle*.[1] The former refers to Mach as an experimentalist, the second – as a philosopher of science and critic of Newton's *Principia*. It is the second aspect that interests us in this paper.

There is no need today to deal with Mach's epistemological doctrine which once was the focus of controversy when Mach's philosophy of science was under discussion.[2] Today, we realize that the doctrine of the 'neutral elements' as expounded in the *Analyse der Empfindungen* is but a miscarried by-product of Mach's endeavour to create, on the basis of historical, methodological and psychological studies, a unifying philosophy of science that would also solve the psycho-physical problem. The centre of interest in Mach's philosophy of science lies not with epistemology or logic of science but with the *historical development of science as a human activity* – a fact that has been implicitly acknowledged by MACH[3] himself, but apparently 'forgotten' by most of his followers.

Oddly enough, it was Mach's comment on Newton's mechanics – and not the pertinent ideas of Riemann – that was seized upon by Einstein as one of his guides to a new theory of gravitation. No wonder that it came as a shock to Einstein when it turned out that his 'Mach Principle' was not satisfied by the new theory. As a consequence the latter was generally held by the experts to have no logical connection at all with Mach's analysis. In the last 10 years more sophisticated interpretations of Mach's ideas have been given and shown to be satisfied by certain cosmological models with closed finite space.[4] This has reopened the question as to the logical relation between Mach's ideas and Einstein's theory.

In the following this question will be commented on within the frame of a wider discussion. Our primary object is to clarify the relations between Newtonian mechanics, the Special Theory[5], and the General Theory.[5] Such a clarification does not involve any particular philosophy of science but it does contain many lessons for it. Moreover, it is a prerequisite for any philosophical discussion of these theories and for any proper judgment on such discussion.

There are two sets of reasons why a reconsideration of the relations between the three theories is necessary. In the first place, it is only now that we can claim to have reached a sufficient understanding of Einstein's theories, free of the vagaries and misconceptions that went into their making and first presentations.[6]

The reasons of the second set are mathematical theorems that have been ignored in previous discussions. One of them is the NOETHER [7] *theorem* which establishes a *relation between spacetime symmetries and conservation laws*, and hence between *kinematics* and *dynamics*. This theorem is of course well known to physicists of today, but only few physicists of the previous generation, and hardly any philosophers of science, seem to have taken notice of it. The second theorem [8] is of still greater importance for our discussion. In its general form it is almost trivial:

> *Any group of transformations defines an equivalence class, namely the class of objects transformed into (mapped onto) each other under the group.*

However, this almost trivial theorem leads to the following consequence:

> (General theorem on kinematic equivalence:) *The invariance group of a given spacetime uniquely determines the class of kinematically equivalent frames.*

The consequences of this theorem will be explored in Part I. Part II compares the full theories. In Part III the results thus obtained are used for a critical examination of Mach's comment on Newton's mechanics.

I. COMPARISON OF THE THREE SPACETIMES

1. *Forms of Presentation of a Physical Theory*

A physical theory (PT) may be presented in any of the following forms:

(a)   customary form: $PT = MF \wedge PI$
(b)   axiomatic form: $PT = \bigwedge_i A_i$
(c)   formalized form: $PT = Sy \wedge Se$

where MF = mathematical formalsim, PI = physical interpretation, A_i = axioms, Sy = syntactic rules, and Se = semantic rules.

While the formalized presentation can easily be obtained[9] from the customary presentation, it is doubtful whether the axiomatic form (b) is always obtainable: it implies that the axioms A_i be such that they (1) need no physical interpretation and (2) determine the mathematical formalism. In spite of this, physical axiomatics has proved extremely valuable in clarifying the physical content of a theory. This applies in particular to Einstein's Special Theory.[10] Unfortunately, no axiomatic presentation of the General Theory is available to which reference could be made.

In the following the customary form is implied.

2. Structure of a Physical Theory

If we write

$$PT = MF \wedge PI \qquad (1)$$

this may be said to give the gross structure of a physical theory.

If we consider the fine structure of a PT we notice that MF consists of (a) the *fundamental equations* (FE) (e.g. Maxwell's equations), (b) the *underlying mathematical calculus or structure* (e.g. Minkowski space), and possibly (c) *restrictions* on the solutions of FE (e.g. boundary conditions at infinity, exclusion of advanced potentials or other causality conditions). FE together with possible restrictions will be called the *mathematical superstructure* (MF^S), the underlying mathematical calculus or structure the *mathematical substructure* (MF_S) of the theory concerned. Hence

$$PT = MF_S \wedge MF^S \wedge PI. \qquad (2)$$

MF^S and PI may be simultaneously changed into MF^{S*} and PI^* in such a way that

$$MF^S \wedge PI \equiv MF^{S*} \wedge PI^* \qquad (3)$$

A trival example would be a regauging of the temperature scale according to

$$\theta^* = \theta_0 \ln \frac{\theta}{\theta_0},$$

which would make the fundamental equations of thermodynamics look more complicated, together with a compensating reinterpretation of the 'temperature' θ^*.

Another almost trivial example is obtained by allowing different units of length and time to be chosen in different inertial frames, whereby the frame-transformation formulae become generalized without changing their physical content if the two new parameters are interpreted as ratios of units.[11]

On the other hand, no example is known where a change in the mathematical substructure can be compensated by a change in physical interpretation, except when substructure and superstructure are identical as in the case of ordinary geometry. But ordinary (3-space) geometry is, strictly speaking, not a self-contained physical theory as all geometrical measurements concern distances between world lines in spacetime. In other words, 3-space geometry is an integral but not a constituent part of spacetime geometry.[12]

3. *Comparison of the Three Spacetimes: Invariants*

Newtonian mechanics and Einstein's theories have one thing in common: their mathematical substructures are 4-dimensional spaces to be interpreted as spacetimes. In this they differ from quantum theories where the mathematical substructure is a Hilbert space. Since the substructure largely predetermines the superstructure, a comparison of the three spacetimes is a prerequisite for a proper understanding of the three theories and their characteristic differences. Moreover, in contrast to 3-space geometry which ignores time, the geometry of spacetimes is a more realistic construct, as is evident from the fact that it implies kinematics.

Any spacetime may be characterized by its topological structure, its invariance group ('Erlanger Program'), or its invariants. The last-named mode of characterization seems nearest to physical thinking since the invariants are supposed to have physical meaning independent of arbitrary conventions. For this reason we shall take this mode of characterization as fundamental.

As to conventions, it should be clearly understood from the outset that in spacetime geometry we are concerned with *two entirely different kinds of conventions.*

The conventions of the first kind apply to all mathematical spaces and concern the *choice of coordinate systems.* Any coordinate system can be thought of as a mapping of the given space onto a real number space of equal dimensions. This mapping is completely arbitrary, apart from being required to be one-one. Hence *all coordinate systems* resulting from one another by different mappings ('point transformations') *are equivalent.* Since the choice of any such system is arbitrary and without physical significance *all physically significant quantities must be invariant under mapping (coordinate) transformations.* It also follows that *all physical equations must be form invariant ('covariant')* under these transformations. This is a completely trivial requirement of exactly the same logical status as the requirement of invariance under change of units, which merely compensates for the arbitrariness in the choice of units.

The conventions of the second kind concern the *choice of frames* (Bezugssysteme); they depend on the theory in question since they reflect the *symmetry properties* and *group structure* of a spacetime. The latter determine whether *global* or only *local* frames exist and which frames are *kinematically equivalent* within the given theory. *Only the choice between kinematically equivalent frames is a convention within the given theory. Without reference to a theory it is meaningless to speak of equivalent frames.*

To avoid confusion, small letters (x) will be used for general (curvilinear) coordinates (which do not have metrical meaning) and capital letters (X) for Cartesian frame coordinates (with metrical meaning). A transition $(X) \to (x)$ or $(x) \to (x')$ is a mere change of coordinates (mappings), without physical significance. A transition $(X) \to (X')$ or $(X^{(i)}) \to (X^{(k)})$ implies a transition from one frame to another one. Similarly, a transition $(dX) \to (dX')$ or $(dX^{(i)}) \to (dX^{(k)})$ implies a transition between local frames.

The two kinds of transitions are completely independent of one another. Their confusion is the root of most misrepresentations and misapprehensions of the General Theory.[13]

Since global frames do not exist in the spacetime of the General Theory (except in special cases), this theory is bound to use general coordinates. On the other hand the use of general coordinates is usually avoided in the presentation of Newtonian mechanics and the Special Theory. This rather obscures the mathematically and physically significant differences. Hence, to facilitate comparison one should use general coordinates for all three theories, in addition to frame coordinates where the latter exist.

The *proper invariant* (frame *and* coordinate independent) quanties are:
(a) in *Newtonian* spacetime:
    (1) global time intervals[14]

$$\varDelta T = \int_{t_1}^{t_2} \gamma_{00}\,(t)\,\mathrm{d}t \tag{4}$$

    (2) global spatial distances

$$|\varDelta L| = {}_+\sqrt{\delta_{ab}\varDelta X^a \varDelta X^b} = \int_{\tau_1}^{\tau_2} {}_+\sqrt{\gamma_{\alpha\beta}\frac{\mathrm{d}x^\alpha}{\mathrm{d}\tau}\frac{\mathrm{d}x^\beta}{\mathrm{d}\tau}}\,\mathrm{d}\tau \tag{5}$$

$(a, b, \alpha, \beta = 1, 2, 3)$;
    (b) in *Minkowski* spacetime: global spacetime intervals

$$|\varDelta S| = {}_+\sqrt{\eta_{mn}\varDelta X^m \varDelta X^n} = \int_{\tau_1}^{\tau_2} {}_+\sqrt{g_{\mu\nu}\frac{\mathrm{d}x^\mu}{\mathrm{d}\tau}\frac{\mathrm{d}x^\nu}{\mathrm{d}\tau}}\,\mathrm{d}\tau \tag{6}$$

$(m, n, \mu, \nu = 0, 1, 2, 3; X^0 = cT; \eta_{00} = 1, \eta_{aa} = -1$, other components $= 0$; τ arbitrary parameter);
    (c) in spacetime of *General Theory*:
        (1) local infinitesimal[15] spacetime intervals

$$|\mathrm{d}S| = {}_+\sqrt{g_{\mu\nu}\,\mathrm{d}x^\mu\,\mathrm{d}x^\nu} = {}_+\sqrt{\eta_{mn}\,\mathrm{d}X^m\,\mathrm{d}X^n} \tag{7}$$

    (2) the scalar curvature field

$$R = g_{\mu\nu}R^{\mu\nu} = \eta_{mn}R^{mn}. \tag{8}$$

Note that *Latin indices* refer to *frames* while *Greek indices* refer to *coordinate systems*.

4. *Comparison of Spacetimes: Group Structure and Kinematically Equivalent Frames*

The proper invariance group of any mathematical space is uniquely determined by the proper invariants of that space.

The proper invariance groups of the spacetimes of the Special and the General Theory are well known: they are the 10-parameter full Lorentz (Poincaré) group L_{10} and the corresponding local (infinitesimal) group δL_{10}. (L_n means an n-parameter Lie group). 4 parameters mean 'translations'[16], i.e., linear displacements $X^m \to X'^m = a_m + X^m$, accounting for the homogeneity of (global or local) spacetime, 3 parameters mean spatial 'rotations'[16], i.e., angular displacements, accounting for (global or local) isotropy of the spacelike hypersurfaces $cT = X^0 = $ const. If these 7 parameters are zero, there remains a 3-parameter subgroup $K = L_3$ or $\delta K = \delta L_3$, respectively, where the 3 parameters mean the 3-vector velocity $\mathbf{v}^{i(k)}$ between frames $\Sigma^{(i)}$ and $\Sigma^{(k)}$. This K (or δK) is the *proper kinematic subgroup* which defines the *family of kinematically equivalent frames*. Thus in Minkowski spacetime the family of kinematically equivalent frames is a so-called *uniform motion equivalence*, i.e., a 3-parameter family of frames $\Sigma^{(i)}$ all in uniform motion with respect to one another. Since the family of kinematically equivalent frames is uniquely determined by the subgroup K, we have

1. THEOREM: *Minkowski spacetime implies the existence of precisely one global uniform motion equivalence.*

In the spacetime of the *General* Theory we have no global symmetries and hence no global invariance groups, but locally the same symmetries and infinitesimal group structure as in Minkowski space. Thus, we have

2. THEOREM: *In the spacetime of the General Theory the one global uniform motion equivalence of Minkowski space is replaced by an infinitude of local uniform motion equivalences.*

What about the group structure and the kinematically equivalent frames in Newtonian spacetime? The answer: it is fundamentally different, since Newtonian spacetime is the *direct product*[17] of time T and space E_3 ($T \times E_3$). For the *non-kinematic subgroup* this does not make much difference: it is again an L_7, only in contrast to the non-kinematic subgroup of Minkowski spacetime it is a direct product: $L_7 = L_1 \times L_6$, corresponding to the *two* kinds of invariants ΔT and $|\Delta L|$, Equations (4) and (5). The *kinematic subgroup* K, however, is not an L_3 but an L_∞. In the first place, instead of the 3-vector $\mathbf{v}$ (const. velocity), we have an arbitrary time function $\mathbf{f}(T)$ as group parameter, corresponding to *arbitrary curvilinear non-rotational motions* between the frames; this gives the *non-rotational subgroup* K_1 ($\mathbf{f}$) of K; it comprises all motions in which the frame axes remain parallel.[18] In the second place, we have the *rotational subgroup K_2* of K, which consists of all possible frame rotations.[19] Thus, the group parameter of K_2 is again an arbitrary time function, or rather a set of such functions: $K_2 = K_2(\omega_b^a(T))$. It follows:

3. THEOREM: *In Newtonian spacetime all frames are kinematically equivalent irrespective of their relative motions.*

This result has of course been assumed or implied by many writers, but it does not follow from the symmetry properties as usually understood [20] (Minkowski space has maximal symmetry, too), but from the 'absoluteness' of time, i.e., the splitting up of the Minkowski invariant $|\Delta S|$, Equation (6), into *two* invariants $|\Delta L|$ and ΔT. Vice versa, *the tremendous reduction of the all-embracing kinematic equivalence class of Newtonian spacetime to a single uniform motion equivalence is solely due to the existence of a finite limiting velocity implied by* (6), which prevents the splitting.

5. *Summary*

The results obtained may be summarized as follows:

(A) Properties of spacetimes.

(1)

	1	2a	2b	3a	3b
	Spacetime	Non-kinematic Symmetries	Kinematic Symmetries	Kinematically Equiv. Frames	Kinematically Distinguished Frames
Newton	$T \times E_3$	7 (global)	∞ (global)	all (global)	none (global)
Einstein I Special Theory	E_{1+3} sign. ± 2	7 (global)	3 (global)	1 global 3-parameter family ('uniform motion equiv.')	
Einstein II General Theory	R_{1+3} sign. ± 2	0 global 7 local	0 global 3 local	(no global frames) ∞ *local* unif. motion equivalences	

E_n = n-dimensional Euclidean space;

E_{n+m} = $(n+m)$-dimensional pseudo-Euclidean space with signature $\pm (n-m)$;

R_{n+m} = $(n+m)$-dimensional Riemannian space with indefinite metric and signature $\pm (n-m)$.

(2) *The geometry of spacelike hypersurfaces* $t = const$ *in Minkowski space is non-Euclidean,* except when $t = T^{(i)}$, i.e., *except in the kinematically distinguished frames of the uniform motion equivalence.*

(B) Relations between spacetimes:

(3) E_{1+3} (*Minkowski space*) is a *special case of* R_{1+3} (*spacetime of the General Theory*), namely the case $R_{\mu\nu\varkappa\lambda} \equiv 0$. These two spacetimes *agree locally.*

(4) $T \times E_3$ (*Newtonian spacetime*) is *not* a special case of E_{1+3} (*Minkowski space*). These two spacetimes *differ even locally in topology and group structure,* though they have the same number (7) of non-kinematic symmetries.

Ad (2): The *physical meaning* of this mathematical fact may be derived as follows. The spacetime defined by light kinematics $[(\Delta S)^2 = \eta_{mn}\Delta X^m \Delta X^n = 0]$ is not Minkowski space but 4-dimensional *conformal* space (15-parameter invariance group).[21] The reduction to Minkowski space (Lorentz group) implies the existence of massive particles, though not necessarily of rigid rods. In addition to 'light geometry' we may thus define a 'particle geometry'. Then: *the preferential frames of Minkowski spacetime* (frames of the one uniform motion equivalence) *are precisely those frames in which particle and light geometry coincide. In all other* (non-global) *frames the two geometries differ.* This may be expressed by saying that in the *non-preferential ('accelerated') frames the propagation of light is anisotropic, judged from the standpoint of particle geometry.* (This statement does not depend on conventions.)

II. PREFERENTIAL FRAMES AND GRAVITATION

6. *Preferential Frames: The First Discrepancy in Newtonian Mechanics and the Leibniz vs. Newton controversy*

If we pass from kinematics to dynamics (equations of motion, field equations) we may or may not encounter a reduction of the class of equivalent frames: kinematically equivalent frames need not be dynamically equivalent. Though there cannot be any logical objection to such a reduction of equivalence, philosophers from Leibniz to Mach have objected to it. Whatever their personal reasons may have been, the objection itself can be justified on sound philosophical and methodological grounds.

If we consider space and time as the forms of existence of matter and not as things in themselves, this implies that *statements about spacetime are indirect statements about the behaviour of matter.* A specific version of this would be: *The properties of spacetime are the properties of properties of matter.* This version is in fact supported by the Noether theorem which connects spacetime symmetries with conservation laws. Thus, the property of *angular momentum* (a property of matter) to be conserved is related to isotropy of space. Now the point (which is sometimes overlooked) is this; the invariance group concerned in the applications of the Noether theorem is not that of spacetime but that of dynamics (equations of motions or field equations). Thus *the properties of spacetime can be interpreted as properties of properties of matter if and only if the spacetime invariance group is identical with* (or a subgroup of)[22] *the dynamical invariance group.* This condition is equivalent to demanding that *all frames kinematically equivalent should also be dynamically equivalent.*

In Newton's mechanics the all-embracing class of kinematically equivalent frames is reduced by the equations of motions to a single uniform motion equivalence, the 3-parameter family of 'inertial frames'. HUYGENS and LEIBNIZ[23] were philosophically right in objecting to this *discrepancy between kinematics and dynamics* but they proved quite wrong in demanding 'general relativity', i.e., dynamical equivalence of all frames. They simply overlooked that the discrepancy may also be removed by *restricting* kinematic equivalence to one uniform motion equivalence instead of extending dynamic equivalence to all frames. We cannot blame Leibniz for not having anticipated Einstein's Special Theory which removes the discrepancy by precisely such a restriction of kinematic equivalence. But we must correct the verdict of REICHENBACH (1924) and other writers who consider Huygens and Leibniz as precursors of Einstein's theories. We even have to admit that Newton's insistence on the existence of preferential frames is fully vindicated[24] by both the Special and the General Theory; the latter merely replaces the one global uniform motion equivalence by a multitude of local uniform motion equivalences, as shown in Part I. The idea that the General Theory is a 'theory of general relativity', meaning equivalence of all frames, is simply due to a mistake in semantics, to wit, the confusion of *frames* and *coordinate systems.* This confusion has first been pointed out in 1917 by KRETSCHMANN (1917), but Einstein's mistaken view prevailed until FOCK (1957, 1960) took the matter up again.

There is no doubt that Einstein's mistaken view in this matter goes back to Mach's influence on the young Einstein, and more specifically to Einstein's acceptance of the doctrine of 'general relativity' (equivalence of all frames) inherent in Mach's comment on Newtonian mechanics, though the doctrine itself was proclaimed, unknown to Einstein, 200 years before by Leibniz. It is an odd but not untypical feature of the history of science that the inventor of the Special Theory did not realize that the *postulate of general relativity* (equivalence of all frames) *implies Newtonian space-time* and hence is *incompatible with the very existence of a finite limiting velocity c.*

7. Preferential Frames: Meaning, Identification, Prediction

In *Newtonian mechanics* the preferential frames are implicitly defined by the equations of motions: this determines their theoretical meaning. But this sort of definition does not allow us to identify a given frame as preferential or non-preferential on mere inspection; in other words, the theory *does not predict* which frames are preferential. Hence, the *identification* has to be done *empirically* or by *further assumptions*.

Moreover, since the kinematic equivalence class of Newtonian space-time contains an infinite number of different uniform motion equivalences, any such identification is a *selection* among an infinite number of equal possibilities.

In the Special Theory the situation is somewhat different. In the first place, the class of preferential frames defined by the equations of motions is *identical* with the kinematic equivalence class; thus *no selection* is implied in the empirical identification. In the second place, the preferential frames of the Special Theory may be defined by the condition that particle and light geometry coincide, which may be called a semi-operational definition. However, the Special Theory does *not predict* which frames are preferential so that here, too, the identification has to be done *empirically* or by *further assumptions*.

On the other hand, the General Theory does predict the local preferential frames as a function of the distribution and motion of matter.

In practice, the preferential frames of Newtonian mechanics have been identified by further assumptions rather than strictly empirically. The procedure of astronomy exemplifies this. The assumptions made always imply that the preferential frames are determined by the distribution and motion of matter in the universe. These assumptions may hence be regarded as a *partial anticipation* of the General Theory. Still, it is this *lack of theoretical* (predictive) *determination*, and not the lack of freedom in the choice of frames ('general relativity') that is a genuine weakness of Newtonian mechanics and, to a lesser degree, of the Special Theory. Curiously enough, it has never been objected to by the critics.

8. *Preferential Frames: Newtonian and Galileian Frames – The Second Discrepancy*

If we come to gravitation, we have to distinguish between Newtonian and Galileian frames. By definition, *Newtonian frames* are those in which Newton's equations of motion (without inertial forces) hold. A test particle in a Newtonian frame would move either with constant velocity or with an acceleration determined by non-inertial forces. A *Galileian frame* is defined by the condition that a test particle moves with constant velocity with respect to the frame, i.e., that the total force is zero. Now the gravitational force is both *universal* and *not to be screened off* (nicht abschirmbar). Hence, *global Galileian frames do not exist in the presence of gravitating matter*.

Here we have a *second discrepancy* in Newtonian theory: what is admitted by its *general dynamics* is excluded by its *gravitational theory*. Oddly enough, this second discrepancy has not been objected to by the critics either.

There exist of course *local Galileian frames* in Newtonian theory. These, and not the Newtonian frames, are the ones that correspond to the local preferential (Minkowski) frames of the General Theory.

9. *Gravitation*

A few remarks only can be made here on the different ways in which gravitation is accounted for in the three theories.

Usually, the General Theory is directly compared with the Newtonian theory of gravitation. Since the mathematical substructures of these two theories are completely different, the comparison must be confined to testable statements. This does not help to clarify the specific roles played by (a) the substructure, (b) the superstructure in producing the differences to Newtonian predictions.

This task would be much easier if we had an acceptable theory of gravitation on the basis of Minkowski spacetime: by writing the gravitational law of such a theory in general coordinates ('general covariance') the comparison could be confined to laws expressed in the same mathematical language; on the other hand, the comparison of such a theory with Newtonian gravitation would also be quite easy.

There exist various ways of getting Lorentz-covariant generalizations of Newtonian gravitation – the simplest one being the introduction of a Lorentz-invariant scalar potential which would account for the retardation of gravitational action. However, none of these possible generalizations is free from serious objections. Hence, in the absence of a recognized Lorentz-covariant theory of gravitation it is best to *define* such a theory as the special case $R_{\mu\nu\varkappa\lambda} \equiv 0$ (Minkowski spacetime) of the General Theory. If this is done, gravitation in the Special Theory is reduced to the so-called inertial forces while the proper gravitational field is zero. This, by the way, shows that inertial and proper gravitational forces are of different standing also in the General Theory – a fact that is merely obscured by general covariance.

There are other ways to facilitate comparison of the General Theory with familiar concepts. One of them, which we shall consider later, is the regaining of the concept of force.

Another one is the introduction of generalized Minkowski frames, i.e., of coordinate systems tending to Minkowski frames for $R_{\mu\nu\varkappa\lambda} \rightarrow 0$. Such coordinate systems are the harmonic coordinates of Fock, defined by $(\sqrt{|g|}\, g^{\mu\nu})_{,\nu} \equiv 0$. These coordinate systems may likewise be used to separate intertial forces from gravitational forces proper and to compare them with the inertial and gravitational forces of Newtonian theory.

This shows that the so-called *principle of equivalence* does not constitute an integral part of the General Theory; in fact it holds only locally. But the same is true in Newtonian theory.

A more comprehensive discussion would show that the differences in physical content between the General Theory and Newtonian theory, as far as gravitation is concerned, result from three distinct sources: (a) retardation of gravitational forces, (b) replacement of scalar potential by tensor potential $g_{\mu\nu}$, (c) chronogeometrical action of matter as implied by interpreting $g_{\mu\nu}$ as metrical tensor to be determined by the field equations. For purely gravitational questions point (a) is decisive; this may be seen from the fact that Newtonian gravitation is the limiting case $c \rightarrow \infty$ of

Einstein's gravitation.[25] Point (c) is decisive for the influence of gravitation on all other physical phenomena; it accounts in a logically most satisfactory way for the two fundamental properties of gravitation: *universality* and *nicht-abschirmbarkeit*. These two properties alone call for a chronogeometrical theory of gravitation if the properties of spacetime are to reflect the properties of properties of matter. But Einstein's theory accounts for an additional property: the *proportionality between inert mass and gravic charge* (gravitational mass) which is not implied by the two other properties.

It would be interesting to know how Einstein's theory had to be modified if inert mass would depend on other factors besides gravic charge or even be independent of it. The answer to this question is not known. All that can be said is this: the field equations for the $g_{\mu\nu}$ would be required *not* to imply the law of motion.

From the logical point of view it may seem surprising that the General Theory, in spite of its entirely different mathematical structure, leads to almost the same predictions as Newtonian theory. But this is a deception: it does so only under two conditions: (a) if the gravitational field is weak (small deviations from Minkowski metric), and (b) if the boundary conditions ensure Minkowski metric at infinity. The latter point concerns cosmology. If solutions representing closed finite spaces are admitted no boundary conditions are required. For sufficiently simple cosmological models it is then even possible to introduce a universal ('absolute') time. This may irritate philosophical relativists; it will not irritate those who consider statements about space and time as indirect statements about matter.

A word of caution should be said about applying Einstein's equations to a fictitious world free of matter $(T_{\mu\nu} \equiv 0)$.

If we were to imagine such a world in Newtonian mechanics the preferential Newtonian frames would become a sort of ghosts: they could not be located and identified. If we turn to the so-called Einstein spaces defined by the condition $T_{\mu\nu} \equiv 0$ we obtain source-free gravitational waves in an otherwise Minkowskian spacetime. There exist other solutions which have been investigated by WHEELER (1963a) in his geometrodynamics. Treatment and physical interpretation are based on an analogy with the Maxwell theory. In the latter the solutions of the source-free (homogeneous) equations have a physical meaning, thanks to the principle of superposition characteristic of linear equations. But Einstein's equations are highly non-linear. For this reason it is not to be expected that the solutions of the homogeneous $(T_{\mu\nu} \equiv 0)$ Einstein equations have any physical meaning.

10. *A Mistaken Criticism of Newtonian Theory*

Frequently Newtonian mechanics is charged with implying fictitious causes for observable phenomena. Some people believe that Mach's analysis has revealed just this. This is not rue: Mach's considerations merely show that the Newtonian inertial frames depend on the distribution of matter in the universe. The only charge that can be derived therefrom is that Newtonian theory does not *predict* what frames are preferential. The mistake in this matter is not due to Mach but to his readers.

The story is different when we pass from Mach's comment on Newton's water pail experiment to Einstein's fictitious experiment with two similar fluid bodies rotating with respect to one another about a common axis. EINSTEIN's (1916) argument rests on the assumption that one of the two bodies is at rest in a preferential Newtonian frame. Then no real cause can be found for the different behaviour of the two bodies.

But *Newtonian mechanics does not imply that the preferential frames can be chosen independently of the distribution and motion of matter*; it just leaves their identification to experience.

In a world consisting of nothing but the two bodies considered by Einstein the preferential frames can even be guessed *a priori* by reasons of symmetry. If this is done, Newtonian theory predicts that *the two bodies will show precisely the same behaviour*, apart from moving towards each other with increasing acceleration due to gravity.

11. *Summary*

(1) *Newtonian theory* contains *three objectionable features*. The first (discrepancy between kinematics and general dynamics) has been removed by the Special Theory, not by extending dynamic equivalence to all frames as demanded by Leibniz but by *restricting* kinematic equivalence to a single uniform motion equivalence. The second (discrepancy between general dynamics and gravitational theory) has been removed by the General Theory, not by generalizing equivalence of frames but by a

further restriction of it to local frames. The third (unpredictability of preferential frames) has likewise been removed by the General Theory, by making the metric $g_{\mu\nu}$, and hence the local Minkowski frames, depend on the distribution and motion of matter $(T_{\mu\nu})$.

(2) *The philosophical merit of the General Theory* lies in the fact that it accounts for the two most fundamental properties of gravitation (universality and nicht-abschirmbarkeit) in a fundamental way, viz., by space-time properties. However, from the physical point of view there is little difference between the local Galilei frames of Newtonian theory and the local Minkowski frames of the General Theory.

The *difference in physical content* between the two theories results in the first place from the *retardation of gravitational action*, already required by the Special Theory, in the second place by the *replacement of a scalar by a 4-tensor potential*, and in the third place by the *non-linearity of the field equations*; the latter is the decisive factor for very strong fields and for cosmology. As a result of the non-linearity, 'matter' and 'gravitation' cannot be separated in the same way as 'electric current' and 'electromagnetic field' in Maxwell's theory; in particular, the gravitational field is interacting with itself.

(3) The *equality of inertial and gravic mass* must be considered a *consequence* rather than a foundation stone of the General Theory: it is a consequence of choosing the *simplest* 4-tensor generalization of Newtonian theory (i.e., of $\nabla^2\chi = 4\pi k\rho$) as field equations; the latter *imply* that the world lines are the same for all test bodies, namely geodesics.

(4) *The heuristic ideas* that guided Einstein in constructing the General Theory *do not form any part of the final theory*.

III. MACH'S COMMENT ON NEWTON'S MECHANICS

12. *The Two Opinions*

There exist two opinions as to the character of Mach's comment on Newtonian mechanics.

The one, proclaimed by Einstein and taken over by most writers on the subject, holds that Mach's comment has revealed fundamental difficulties in the conceptual structure of Newtonian mechanics and thus prepared the way to Einstein's General Theory. The other opinion, held by Mach himself, holds that Mach was criticising Newton's *presentation* of the theory rather than the theory itself. That this was indeed MACH's opinion emerges from the concluding paragraph 6 which reads as follows:

6. Im Ganzen kann man sagen, daß Newton in vorzüglicher Weise die Begriffe und Sätze herausgefunden hat, welche genügend gesichert waren, um auf dieselben weiter zu bauen. Er dürfte zum Teil durch die Schwierigkeit und Neuheit des Gegenstandes seinen Zeitgenossen gegenüber zu einer großen Breite und dadurch zu einer gewissen Zerrissenheit der Darstellung genötigt gewesen sein, infolge welcher z.B. ein und dieselbe Eigenschaft der mechanischen Vorgänge mehrmals formuliert erscheint. Teilweise war er aber nachweislich über die Bedeutung und namentlich über die Erkenntnisquelle seiner Sätze selbst nicht vollkommen klar. Und auch dies vermag nicht den leisesten Schatten auf seine geistige Größe zu werfen. Derjenige, welcher einen neuen Standpunkt zu erwerben hat, kann denselben natürlich nicht von vornherein so sicher innehaben, wie jene welche diesen Standpunkt mühelos von ihm übernehmen. Er hat genug getan, wenn er Wahrheiten gefunden hat, auf die man weiter bauen kann. ... Später wird dies anders. Von den beiden folgenden Jahrhunderten durfte Newton wohl erwarten, daß sie die Grundlage des von ihm Geschaffenen weiter untersuchen und befestigen (!) würden. ... Dann treten Fragen auf, wie die hier behandelten, zu deren Beantwortung hier vielleicht ein kleiner Beitrag geliefert worden ist. ..." [26]

Still, it will be necessary to take a closer look at Mach's comment in order to show that the non-Machian opinion on it is wholly mistaken. As a by-product it will turn out that part of Mach's comment is itself in need of criticism.

13. *Metric of Newtonian Time*

Newton's equations of motion are invariant under a linear, but not invariant under a non-linear transformation of the time variable. A linear transformation

$$T' = aT + b \tag{9}$$

does not affect the *metric* of the time scale, i.e., the definition of equality of time intervals; indeed, (9) implies

$$T_4' - T_3' = T_2' - T_1' \leftrightarrow T_4 - T_3 = T_2 - T_1 \tag{10}$$

and vice versa. Thus, *the metric of Newtonian time is implicitly defined by the equations of motion.*

Newton was well aware of the *practical* need to find a real standard of time that would satisfy the implicit definition. But he also knew that the conventional standards then used (astronomical clocks) did not fully satisfy this requirement. Rather than to subscribe to any arbitrary choice he took the implicit definition as the standard definition. This, and nothing else, is implied in what Newton calls the "absolute, true, mathematical" time. Unfortunately, he starts his introduction of this concept with the nonsensical assertion that this time flows uniformly and without regard to anything external. Only afterwards when he speaks of the conventional standards of time does it emerge what he really had in mind.

For his nonsensical explanation Newton has earned all the criticism levelled on it. But this criticism concerns Newton's way of presenting the theory; it does not in the least alter the crucial fact that the metric of Newtonian time is uniquely defined by the equations of motion and that this time may rightly be called 'absolute, true, mathematical'.

Conventionalists may argue that a non-Newtonian time scale may also be used even though it would lead to mathematical and semantic complications in the statement of the law of motion. Yet the new time scale would still be defined implicitly by the equations of motion; only we would not find any natural clock keeping approximately non-Newtonian time.

Newton's insistence on the preferential status of his time scale is fully vindicated by the modern (Hamiltonian) form of mechanics, both classical and quantum: *Newtonian time is identical with the canonical time τ* defined by the condition

$$\Omega(\tau_2)\,\Omega(\tau_1) = \Omega(\tau_2 + \tau_1) \tag{11}$$

where $\Omega(\tau)$ is the operator of motion in state space (Zustandsraum) defined by

$$P(\tau + \tau_0) = \Omega(\tau)\,P(\tau_0), \tag{12}$$

$P(\tau)$ representing the state of the system at time τ. Moreover, unless τ_0 is a distinguished instant of time (and hence to be treated as a constant instead of as a variable), (12) implies (11). Thus, *canonical time and Newtonian time are identical*.

Mach's comment on Newtonian time runs over three pages, but it does not emerge from it whether Mach has even understood the question at issue, viz., the problem of defining a time metric. It does seem, however, that he would deny the possibility of a unique theoretical definition of time metric, since he writes that the concept of time must be based on the comparison of changes and that absolute time "ist ein müßiger, metaphysischer' Begriff".[27] If 'absolute time' means 'Newtonian time' this is just as wrong as a corresponding statement about 'absolute temperature' would be. If he does not mean Newtonian time but Newton's introductory statement on it, he is merely criticising Newton's *presentation* of the theory.

14. *Frame Time, Universal Time, Simultaneity*

Nowadays the expression 'absolute time' is usually employed to denote what should be called *universal time*, in contrast to frame time (extended local time).

Newtonian time is of course universal, the same for all frames.

There is no indication in Mach's comment that he is objecting to a universal time. In spite of this, Mach is often credited with having inspired Einstein's Special Theory. The connecting link is seen in Einstein's analysis of the concept of simultaneity which is said to be conducted in the spirit of Machian empiricism. Einstein has acknowledged this when he wrote

Das kritische Denken, dessen es zur Auffindung dieses zentralen Punktes bedurfte, wurde bei mir entscheidend gefördert insbesondere durch die Lektüre von *David Humes* und *Ernst Machs* philosophischen Schriften.[28]

But he also wrote

Beim Fehlen dieser aus der Maxwell-Lorentz'schen Elektrodynamik fließenden Anregung [Konstanz der Lichtgeschwindigkeit, M.S.] reichte auch Machs kritisches Bedürfnis nicht hin, um das Gefühl der Notwendigkeit einer Definition der Gleichzeitigkeit örtlich distanter Ereignisse zu wecken.[29]

The two points that emerge are these:

(1) In Einstein's opinion the necessity of explicitly defining simultaneity is a central point;

(2) Mach has missed this point because the implications of Maxwell's electrodynamics were not taken seriously.

The true story is far more complicated and deserves a separate study. Here I confine myself to three remarks.

First, the existence of infinite signal velocities, assumed in Newtonian kinematics, is sufficient but *not necessary* to establish a universal time. If the universe has a natural history the latter would define a universal time. This need not conflict with the Lorentz group; a possible model theory of this kind has been given by MILNE[30]. Some cosmologies of the General Theory also admit a universal time without implying infinite velocities.

Second, the significance of Einstein's analysis of simultaneity is vastly overrated both by himself and by many philosophers. In the first place, it is based on the conventional definition of velocity, viz.,

$$
\left. \begin{array}{l}
\mathbf{v}_k^i \underset{Df}{=} \left[\dfrac{d\mathbf{r}^{(i)}}{dt^{(i)}} \right]_{d\mathbf{r}^{(k)} = 0} \\[2em]
\mathbf{v}_i^k \underset{Df}{=} \left[\dfrac{d\mathbf{r}^{(k)}}{dt^{(k)}} \right]_{d\mathbf{r}^{(i)} = 0},
\end{array} \right\}
\tag{13}
$$

which leads to *two* independent mathematical expressions for one and the same physical relation – a logical absurdity that has to be corrected for by a separate postulate such as

$$
|\mathbf{v}_k^i| = |\mathbf{v}_i^k| \tag{14}
$$

or

$$
v_k^i = - v_i^k. \tag{15}
$$

In the second place, Einstein's definition of simultaneity is merely *one* of a number of admissible operational definitions compatible with the implicit definition of frame time as given by the Lorentz group. In particular, it is *not a convention* that could be replaced by a different (inequivalent) one as held by Reichenbach and others.

Third, Mach demands that 'time' be derived from the comparison of changes of real objects. Most likely, this implies a dynamic rather than a kinematic conception of time. However, even if we remain within the realm of kinematics it would demand that we should consider *velocity* – a relation between two things – as a *primary concept* and 'time' as a secondary or derived concept defined by (13). Such an approach leads immediately to relation theoretical questions such as to the composition law for constant relative velocities which has been answered more than 100 years ago by Fizeau's experiments. Such a realistic approach, carried through [31] for the first time in 1957, is much more in line with Mach's methodological ideas than Einstein's analysis which appears as a half-way house between the conventional ideas and the ideas of Mach. It is also more in line with Mach's ideas than the actual procedure used by Einstein for obtaining the Lorentz transformation in which his analysis of simultaneity plays no role whatsoever, viz., the establishment of the Lorentz transformation as a *sufficient* condition for the compatibility of two physical principles. This method, ingenious though it was, does not establish the Lorentz group as the invariance group of all physics.

Thus, there is no need at all to invoke external influences for explaining Mach's negative attitude towards Einstein's theory.

15. *Space, Motion, Gravitation*

MACH[32] introduces his comment on Newton's views about space and motion with the statement "Ähnliche Ansichten wie über die Zeit entwickelt Newton über den Raum und die Bewegung". This makes it clear that what Mach is criticising is indeed Newton's views, not Newtonian theory. But even so it obscures the fact that Newton's own concept of absolute time is but a metaphorical transcription of a proper physical concept, viz., Newtonian time as defined by the equations of motion, while there is nothing at all in Newtonian theory that corresponds to Newton's own concept of absolute space; as already pointed out, this concept is not only 'metaphysical' but inconsistent since it implies both homogeneity and inhomogeneity. Only a reinterpretation of 'absolute space' as 'uniquely determined preferential frame' would give a consistent concept. But what Newtonian dynamics does imply is not the existence of such a frame but the existence of a 3-parameter family of equivalent preferential frames while Newtonian kinematics implies no preferential frames at all.

Whether Mach has even noticed this discrepancy between Newtonian kinematics and dynamics does not emerge from his comment; he merely insists on the analytic statement that motion is a relation between things, the things being either bodies or a body and a medium. This statement implies no objection against the existence of preferential frames which, even if not real bodies, are somehow determined by the distribution and relative motion of real bodies. If Mach has contributed to make physicists realize this implication of Newtonian mechanics this may be called a *correction of wrong views* (including the views of Newton) *on Newtonian mechanics*, but not a criticism of Newtonian theory. The best that can be said on his modified water pail experiment is that it leads to a *prediction of* the *preferential frames* and thus fills a gap in Newtonian theory which leaves the identification of the preferential frames to experience. But practical astronomy has always filled this gap in just the same way when it anchored one of the preferential frames in the so-called fixed stars. Indeed, Mach himself acknowledges this; he writes:

Als nun Newton die seit Galilei gefundenen mechanischen Prinzipien auf das Planeten-
system anzuwenden suchte, bemerkte er, daß, soweit dies überhaupt beurteilt werden kann,
die Planeten gegen die sehr entfernten, scheinbar gegeneinander festliegenden Weltkörper,
von Kraftwirkungen abgesehen, ebenso ihre Richtung und Geschwindigkeit beizubehalten
scheinen, als die auf der Erde bewegten Körper gegen die festliegenden Objekte der Erde.
Das Verhalten der irdischen Körper gegen die Erde läßt sich auf deren Verhalten gegen die
fernen Himmelskörper zurückführen. ... Wenn wir daher sagen, daß ein Körper seine
Richtung und Geschwindigkeit im Raume beibehält, so liegt darin nur eine kurze An-
weisung auf die Beachtung der ganzen Welt.[33]

Thus Mach is fully aware that he is but explicating the empirical content of Newtonian theory, in particular that of the concept of preferential
(inertial) frames. It was Einstein who, by an act of ingenious misunderstanding, promoted this explicatory comment to the status of a new
physical principle, called 'Machsches Prinzip', by projecting into it his
own ideas.

16. *Mach's Empiricist Transcription of Newtonian Theory: Elimination of Frames and Forces*

According to Mach, frames and forces are but auxiliary concepts that
should be eliminated from the equations of motion if the empirical content of the latter is to be brought to light. Mach has done certain steps
in this direction, but it is not at all difficult to carry out the program in
full generality for point masses and charges.

In an arbitrary frame Σ^0 we have Newton's equations

$$\frac{d^2 \mathbf{r}_{i0}}{dt^2} = \mathbf{a}_{i0} = \mathbf{f}_i / M_i, \tag{16}$$

where $\mathbf{f}_i$ is the total force on a point mass M_i which, for arbitrary Σ^0,
will include the so-called inertial or apparent forces. On eliminating Σ^0
we have

$$\mathbf{a}_{ik} = \mathbf{a}_{i0} + \mathbf{a}_{0k} = \mathbf{a}_{i0} - \mathbf{a}_{k0} = \frac{M_k \mathbf{f}_i - M_i \mathbf{f}_k}{M_i M_k}. \tag{17}$$

Now the forces $\mathbf{f}_i$ can, according to Mach, only be due to the presence
of the other masses M_i and charges Q_l:

$$\mathbf{f}_i = \Sigma' \mathbf{f}_{il}^{(M)} + \Sigma' f_{il}^{(Q)}, \tag{18}$$

where $\mathbf{f}_{il}^{(M)}$ and $\mathbf{f}_{il}^{(Q)}$ can only depend on M_i, M_l, (or Q_i, Q_l), their distance
$\mathbf{r}_{il}$ and the time derivatives of $\mathbf{r}_{il}$:

$$\mathbf{f}_{il}^{(M)} = \mathbf{f}^{(M)}(M_i, M_l, \mathbf{r}_{il}, \dot{\mathbf{r}}_{il}, \ldots) \tag{19}$$

$$\mathbf{f}_{il}^{(Q)} = \mathbf{f}^{(Q)}(Q_i,\, Q_l,\, \mathbf{r}_{il},\, \dot{\mathbf{r}}_{il},\, \ldots). \tag{20}$$

If we confine ourselves to Newtonian gravitation we have

$$\mathbf{f}^Q = 0, \quad \mathbf{f}_{il}^M = G\,\frac{M_i M_l}{r_{il}^3}\,\mathbf{r}_{il} \tag{21}$$

which yields

$$\frac{d^2 \mathbf{r}_{ik}}{dt^2} = \mathbf{a}_{ik} = G\left[\frac{M_k + M_i}{(r_{ik})^3}\,\mathbf{r}_{ik} + \Sigma'' M_l\left(\frac{\mathbf{r}_{il}}{(r_{il})^3} - \frac{\mathbf{r}_{kl}}{(r_{kl})^3}\right)\right]. \tag{22}$$

This represents the Newtonian law of gravitational motion in the Mach transcription. Its two most important properties: (1) invariance under the *full* invariance group of Newtonian spacetime ('general relativity') and (2) interpretation of inertial forces in terms of Newtonian gravitation of all masses of the universe [second expression in (22)].

Quite obviously, it was this combination of properties that inspired Einstein's hope to obtain a similar result in field theory. The crucial point overlooked by Einstein was this: 'general relativity' is only obtainable in Newtonian spacetime, as shown above (Part I). Moreover, while Mach explains inertial forces in terms of true gravitational forces, Einstein originally thought he could explain true gravitational forces in terms of inertial forces, i.e., as purely kinematic effects.

What is not obvious but curious is that anybody could mistake Mach's transcription for a criticism of Newtonian theory, while *de facto* it was the precise opposite: it portended to show that this theory is in full accord with the requirements of physical empiricism and that the (still popular!) opinion Newtonian theory implies a metaphysical action of space on matter is wholly mistaken.

The question which neither Mach nor any of his followers bothered to investigate is this: *how much physical content is lost by the Mach transcription?* If gravitational motion only is considered it may appear that physical content is gained rather than lost, provided the Σ''-term in (22) is accepted as a correct account of what in the usual version of the theory is called inertial forces. But it does not appear that the transcribed equations are sufficient for solving collision problems. We cannot regain the original equations once the frame of reference is eliminated, and this entails the loss of the greater part of ordinary dynamics. The same conclusion is reached by the following consideration. According to Newton's equations the mathematical expression for the force f_{i0} must be invariant under the proper Galilei group; this implies that it does not involve the velocity of the particle relative to the frame. It follows that the force f_{ik} between two particles must be independent of the relative velocity $v_{ik} = \dot{r}_{ik}$. *But no such restriction is implied if we consider the transcribed equations as fundamental.* Thus, there is a definite loss of content. Even if we were to correct for this we would still find that the *transcribed equations contain only that part of physical content of Newtonian theory that can be expressed in terms of trivial invariants*, i.e., the relational quantities r_{ik}, $\dot{r}_{ik}$, ... that are invariants by definition.

If we admit mass points carrying electric charges Q_l we know from experience that f_{ik}^Q depends on both r_{ik} and $\dot{r}_{ik}$. The Mach equations would allow for this while Newton's equations do not. Thus, it may appear that the Mach equations have a wider scope. But this is not so: electrodynamics, even if we disregard field theory, cannot be expressed within the frame of the Mach equations because it involves an invariant velocity c incompatible with the kinematic substructure of both Newton's and Mach's equations.[34]

Thus, Mach's empirist transcription, while leading to an interpretation of inertial forces in terms of true gravitational forces, is connected with a loss of physical content in general dynamics without affording a true generalization.

17. *The Story of the 'Mach Principle'*

As shown above, there is nothing in Mach's comment on Newtonian mechanics that could be construed as a physical or epistemological principle demanding a new theory. Neither the first nor the second discrepancy of Newtonian theory nor the contingent equality of inert and gravic mass have been objected to by Mach. What he did object to were the objectionable features in Newton's *presentation* of the theory. There is only one point where Mach went beyond Newton: in the *interpretation* of the *inertial forces* as true gravitational forces due to distant masses; but even with this interpretation he wholly remained within the frame set by Newtonian theory. Hence it seems clear from the very beginning that there cannot be any logical connection between Mach's comment and Einstein's General Theory. However, Mach's comment concerns Newtonian mechanics and not the mechanics of Einstein's Special Theory which was the point of departure for the General Theory. Hence we should first answer the question whether Mach's interpretation of the inertial forces can be carried over into the Special Theory; if it cannot, the Special Theory would be less satisfactory from the Machian point of view than Newtonian mechanics.

Now the answer to this question is (essentially) in the negative. The reason is not that we cannot formulate a Lorentz covariant theory of gravitation – that we *can* do – but that we *cannot get rid of the preferential frames defined by the Lorentz group. This changed situation*, and not Mach's comment on Newtonian mechanics as such, must be taken as the rational element in Einstein's heuristic arguments for a generalization of the Special Theory, and especially for the pronouncement of his 'Mach Principle'.

I said 'rational', not 'correct': the 'Mach Principle' as formulated by EINSTEIN [35] is *not* satisfied by the General Theory. This has often been taken as an argument against Mach. But this is not correct either: Einstein's 'Mach Principle', far from being a faithful rendering of Mach's basic idea concerning inertial forces, is a far stronger principle which synthesizes a purely chronogeometrical conception of gravitation with a number of other ideas; more likely than not the principle is 'superstrong', viz., inconsistent.

I now come to a different formulation of Mach's ideas in terms of Riemannian field theory given in recent years by HÖNL.[36] This formulation differs from Einstein's 'Mach Principle' in two fundamental points. In the first place, the concept of *force* is re-established within the General Theory. In the second place, to the masses of the universe, now represented by the matter tensor $T_{\mu\nu}$, is added the energy-momentum density of the gravitational field. That the two *together uniquely* determine the force on a test particle is the demand of Hönl's modified Mach Principle. It is then shown that this principle is satisfied by those cosmological solutions of the Einstein equations that represent closed finite spaces, but (in general) not satisfied by solutions representing open spaces.

This is a rather far cry from the simple considerations of Mach presented above, and I must leave it to the reader to recognize, or not to recognize, the Hönl-Mach Principle as an adequate rendering of Machian ideas. Whatever the verdict may be it does not change the fact that the

Einstein equations themselves have no logical connection with Mach's comment on Newtonian mechanics.

18. *Einstein's First Interpretation of Mach's Comment and the Mass Problem*

Einstein's 'Mach Principle' was proclaimed in 1918, so to speak *post festum*. It was not his first interpretation of Mach's comment and not the one that guided his first steps. What Einstein had hoped for was a theory explaining *inertia as a result of gravitational interaction*. In the light of the facts this idea can only be called a curious misinterpretation of Mach's comment.[37] There is nothing in Mach's comment that would suggest a reduction of inert mass, as distinct from inertial forces, to more fundamental concepts. There is nothing in Einstein's theory either that would allow us to conceive inert mass as a result of mutual interaction. If Einstein has made an advance towards explaining inert mass he has done so by his Special Theory, not by his General Theory. But even in the Special Theory 'rest mass' and 'rest energy' are merely synonyms. The problem of explaining the mass ratio of elementary particles remained completely unsolved. It is now in the process of being solved, but the results obtained are based on quantum and group theory and have nothing to do with gravitation. The most that can be hoped for from gravitational theory as a contribution to a theory of mass is the determination of a *common cosmological factor* in the masses of the elementary particles.[38]

19. *Summary*

Mach's influence on the creator of the General Theory must be viewed as a historical coincidence with positive and negative consequences none of which have a direct logical connection with Mach's comment. Both the positive and the negative consequences are due to a 'free interpretation' of Mach's comment that amounts, logically speaking, to a misinterpretation. The original misinterpretation is the idea that inert mass should be due to gravitational interaction – an idea that is, and was, wholly irrational in view of the facts of atomic physics. The remaining misinterpretation is the 'Mach Principle' of 1918 together with the confusion of general covariance and general relativity. These misinterpretations have obscured the true physical meaning of the theory for a rather long time to all but the experts. This would not have happened if

Einstein would have been guided by the pertinent ideas of Riemann. On the other hand, Einstein's mistaken belief to satisfy Machian epistemological principles was no doubt a driving force that kept him working on a problem that otherwise would not have been solved this way until very much later, viz., until all possibilities of a gravitational theory within the compass of the Special Theory had been exhausted and found wanting.

If one compares the genesis of the General Theory with that of Planck's quantum theory[39] one is struck both by a curious parallelism and by a characteristic contrast. The driving force in Planck's work was an idea no less mistaken than the original ideas of Einstein, viz., the idea that thermodynamics and Maxwell's electrodynamics could be unified into a consistent theory which was to crown the edifice of classical theory. Planck, a revolutionary against will and conviction, vastly underestimated the revolutionary implications of his work and did not accept them when they became apparent. Einstein, on the other hand, overestimated[40] the revolutionary character of his General Theory which to the present generation would appear as the crowning of classical physics if it would not have opened the door to new perspectives in cosmology.

APPENDIX

1. *Transformation Group and Equivalence*

Let $\{e_i\} = E$ be a set of elements and $\{0^{(ik)}\} = \Omega$ a set of operators such that

$$e_i = 0^{(ik)} e_k.$$

Then

$$0^{(ii)} = I$$
$$0^{(ki)} = [0^{(ik)}]^{-1}$$
$$0^{(ik)}0^{(kl)} = 0^{(il)}.$$

Thus the set Ω is a group of transformations in E.

If we define the relation $\overset{\Omega}{\sim}$ by

$$e \overset{\Omega}{\underset{df}{\sim}} e' = \exists 0 \, (0\varepsilon\Omega \wedge e = 0e')$$

it follows from the group properties that $\overset{\Omega}{\sim}$ is an equivalence relation.

2. Galilei and Lorentz Group as Representations of the Velocity Group

Transformation groups in linear vector spaces are usually considered as representations of algebraic or abstract groups, the vector space then being called representation space. The algebraic group underlying the Galilei and Lorentz transformation groups is the velocity group characterized by

$$\mathbf{v}_k^i \leftrightarrow \mathbf{v}_l^k = \mathbf{v}_l^i.$$

The unit element of the group is

$$\mathbf{v}_l^i = \mathbf{v}_k^k = \cdots = \mathbf{0}$$

and the inverse to $\mathbf{v}_k^i$ is $\mathbf{v}_i^k$:

$$\mathbf{v}_k^i \leftrightarrow \mathbf{v}_i^k = \mathbf{0}.$$

The proper Galilei group is a *reducible* true representation of the velocity group in 4-space, the time T being an invariant subspace. The proper Lorentz group is an *irreducible* true representation of the velocity group in 4-space. (There exist other irreducible representations, e.g., by unitary transformations in Hilbert space; these are used in quantum field theory.)

3. The Uniqueness Theorem

Consider two disjunct sets $\{e_i\}$ and $\{\eta_\mu\}$ and assume (notations as under 1)

$$e_i \overset{\Omega}{\sim} e_k, \qquad \eta_\mu \overset{\Omega}{\sim} \eta_\nu.$$

Define the operator set $\{\theta^{k\mu}\}$ by

$$e_k = \theta^{k\mu}\eta_\mu.$$

Then

$$0^{ik}\theta^{k\mu} = \theta^{i\lambda}0^{\lambda\mu} = \theta^{i\mu}. \tag{23}$$

If Ω is a reducible representation no conclusion can be drawn as to the character of the θ. If Ω is an irreducible representation, the last set of equations admits two solutions: *either*

$$\theta^{k\mu} = 0^{k\mu}, \quad \text{i.e.,} \quad \{e_k\} \overset{\Omega}{\sim} \{\eta_\mu\} \tag{24a}$$

i.e., the two sets belong in fact to the same equivalence class; *or*

$$0^{ik} = \theta^{i\mu} (\theta^{k\mu})^{-1} \\ 0^{\mu\nu} = (\theta^{i\mu})^{-1} \theta^{i\nu} \quad (\theta^{k\mu} \neq 0^{k\mu}) \Big\}.$$ (24b)

However, in the second case Ω is not a true representation, i.e. the matrices 0^{ik} would depend on additional parameters (an example of such a non-true representation of the velocity group has been discussed in STRAUSS, 1965).

4. *A Sufficient Condition for the Existence of Several Uniform Motion Equivalences*

Let $\{e_i\}$ be a uniform motion equivalence belonging to the group Ω, this group being either the proper Galilei or the proper Lorentz group.

Define a new set of frames by

$$\eta_i = \theta e_i.$$

Then the necessary and sufficient condition for $\{\eta_i\}$ to be a uniform motion equivalence belonging to the same group Ω is

$$\theta 0^{ik} = 0^{ik}\theta$$ (25)

which is a special case of Equation (23). If the 0^{ik} are the matrices of the Galilei group the possible solutions for θ have been given in the text: there exist an infinite number of solutions. If the $\{0^{ik}\}$ are an irreducible representation (Lorentz group) it follows from Schur's lemma that

$$\theta = \text{const. } I$$

(*I* being the identity operator). Thus, there is no possibility to generate a second uniform motion equivalence related to the first by application of the same operator, except in the case of the Galilei group.

Though the condition (25) has not been proved to be also a *necessary* condition for the existence of a second uniform motion equivalence the result shows clearly that the existence of different uniform motion equivalences in accelerated motion with respect to one another (which is the basis of the doctrine of general relativity) is restricted to the Galilei group (reducible representation), i.e., to Newtonian spacetime.

BIBLIOGRAPHY

Alexander, H. G.: 1956, *The Leibniz-Clarke Correspondence*, Manchester.
Bateman, H.: 1909/10, 'The Transformation of the Electrodynamic Equations', *Proc. London Math. Soc.*, ii. ser., **8**, 223–264.
Casmore, D. C.: 1963, 'Integrable Transformations between Moving Observers', *Proc. Phys. Soc.* **81**, 181–185.
Cassirer, E. (ed.): 1904, *G. W. Leibniz – Hauptschriften zur Grundlegung der Philosophie*, Leipzig, pp. 242–245.
Cunningham, E.: 1909/10, 'The Principle of Relativity in Electrodynamics and an Extension thereof', *Proc. London Math. Soc.*, ii. ser., **8**, 77–98.
Dautcourt, G.: 1964, 'Die Newtonische Gravitationstheorie als strenger Grenzfall der Allgemeinen Relativitätstheorie', *Acta Physica Polonica* **25**, 637–647.
Einstein, A.: 1916a, 'Die Grundlage der allgemeinen Relativitätstheorie', *Ann. d. Phys.* **49**, 771–814.
Einstein, A.: 1916b, 'Ernst Mach', *Phys. Z.* **17**, 101–104.
Einstein, A.: 1918, 'Prinzipielles zur allgemeinen Relativitätstheorie', *Ann. d. Phys.* **55**, 241.
Fock, V. A.: 1957, 'Three Lectures on Relativity Theory', *Rev. Mod. Phys.* **29**, 325.
Fock, V. A.: 1960, *Theorie von Raum, Zeit und Gravitation*, Berlin.
Gürsey, F.: 1963, 'Reformulation of General Relativity in Accordance with Mach's Principle', *Ann. of Phys.* **24**, 211–242.
Heller, K. D.: 1964, *Ernst Mach*, Wien-New York.
Herneck, F.: 1966, 'Die Beziehungen zwischen Einstein und Mach, dokumentarisch dargestellt', *Wiss. Z. Friedrich-Schiller-Univ. Jena, Math.-Nat. R.*, Heft 1, **15**, 1–14.
Hönl, H.: 1966a, 'Zur Geschichte des Machschen Prinzips', *Wiss. Z. Friedrich-Schiller-Univ. Jena, Math.-Nat. R.*, **15**, 25–36.
Hönl, H.: 1966b, 'Das Machsche Prinzip und seine Beziehung zur Gravitationstheorie Einsteins', in *Einstein-Symposium* (ed. by H.-J. Treder), Berlin, pp. 238–278.
Kretschmann, E.: 1917, 'Über den physikalischen Sinn der Relativitätspostulate', *Ann. d. Phys.* **53**, 575.
Leibniz, G. W.: 1691–95, Briefwechsel mit Huygens, s. Cassirer (1904).
Lenin, W. I.: 1909, *Materialismus und Empiriokritizismus*, Moskau. (In Russian.)
Lubkin, E.: 1961, *Frames and Lorentz Invariance in General Relativity*, Univ. of California, UCRL 9668 Internal, April 19.
Mach, E.: 1883, *Die Mechanik in ihrer Entwicklung, historisch-kritsch dargestellt*, Leipzig.
Mach, E.: 1917, *Erkenntnis und Irrtum*, 3rd ed., Leipzig.
Magie, W. F.: 1935, *A Source Book in Physics*, New York-London.
Milne, E. A.: 1948, *Kinematic Relativity*, Oxford.
Noether, E.: 1918, 'Invariante Variationsprobleme', *Göttinger Nachr.*, 235–257.
Pirani, F. A. E.: 1957, 'Tetrad Formulation of General Relativity Theory', *Bull. Acad. Polon. Sci.* **5**, 143–147.
Planck, M.: 1910, 'Zur Machschen Theorie der physikalischen Erkenntnis', *Phys. Z.* **11**, 1186.
Reichenbach, H.: 1924, 'Die Bewegungslehre bei Newton, Leibniz und Huygens', *Kantstudien* **29**, 416.
Schilpp, P. A. (ed.): 1949, *Albert Einstein: Philosopher-Scientist*, Evanston, Ill.
Strauss, M.: 1938, 'Mathematics as Logical Syntax – A Method to Formalize the

Language of a Physical Theory', *J. of Unified Science (Erkenntnis)* 7; [Chapter VI of this volume].

Strauss, M.: 1957/58, 'Grundlagen der Kinematik, I – Die Lösungen des kinematischen Transformationsproblems', *Wiss. Z. Humboldt-Univ.-Berlin, Math.-Nat. R.*, 7, 609–616.

Strauss, M.: 1960, 'Max Planck und die Entstehung der Quantentheorie', in *Forschen und Wirken, Festschrift zur 150-Jahr-Feier d. Humboldt-Univ. Berlin*, I, Berlin, pp. 367–399; [Chapter IV of this volume].

Strauss, M.: 1965, 'On the Voigt-Palacios-Gordon Transformation and the Kinematics Implied by it', *Nuovo Cim.* (X) **39**, 658-666.

Strauss, M.: 1966, 'The Lorentz Group: Axiomatics – Generalizations – Alternatives'. *Wiss. Z. Friedrich-Schiller-Univ. Jena, Math.-Nat. R.* **15**, 109–118; [Chapter XIV of this volume].

Strauss, M.: 1967a, 'Zur Logik der Begriffe "Inertialsystem" und "Masse"', in *Ernst-Mach-Symposium 1966 in Freiburg* (in press); [Chapter XIII of this volume].

Strauss, M.: 1967b, 'Grundlagen der modernen Physik', in *Mikrokosmos – Makrokosmos*, II (ed. by H. Ley), Berlin 1967.

Treder, H.-J.: 1966, 'Lorentz-Gruppe, Einstein-Gruppe und Raum-Zeit-Struktur', in *Einstein-Symposium* (ed. by H.-J. Treder), Berlin, pp. 57–75.

Wheeler, J. A.: 1963a, 'The Universe in the Light of General Relativity', in *Lectures in Theoretical Physics*, V (ed. by W. E. Brittin, B. W. Downs, and J. Downs), New York-London-Sydney, pp. 504–527.

Wheeler, J. A.: 1963b, 'Mach's Principle as Boundary Condition for Einstein's Field Equations', in *Lectures in Theoretical Physics*, V (ed. by W. E. Brittin, B. W. Downs, and J. Downs), New York-London-Sydney, pp. 528–578.

NOTES

* Reprinted from *Synthese* **18** (1968) 251–84. [This article forms Part II of 'Intertheory Relations'.]

[1] The expression 'Machsches Prinzip' is due to Einstein (1918) and denotes an interpretation of Mach's comment on Newtonian mechanics in terms of Riemannian field theory: cf. note 35.

[2] Cf. e.g. Lenin (1909) and Planck (1910).

[3] Cf. the review article on Mach's *Erkenntnis und Irrtum* by F. Jodl, republished as Appendix to this work on the direction of Mach. – Mach (1917), pp. 464–470.

[4] Hönl (1966a, b), Gürsey (1963), Wheeler (1963b).

[5] These names are used for what is commonly called 'Special Theory of Relativity' and 'General Theory of Relativity', respectively. These and similar names involving 'relativity' are misleading, as is now generally recognized. The novel name '(Einstein's) Theory of Gravitation' advocated by Fock does not show that the General Theory contains more than a theory of gravitation: *the General Theory generalizes the Special Theory in such a way that a theory of gravitation is included.*

[6] This is no blemish on the theories' inventor: the history of science knows no instance of a physical theory correctly understood by its author. The correct meaning of a new fundamental physical theory only emerges in a long and difficult process of logico-mathematical analysis and practical applications.

[7] Noether (1918).

[8] For proofs cf. Appendix.

[9] Cf. Strauss (1938).
[10] Cf. Strauss (1966).
[11] Cf. M. Strauss (1966), section 4.

[12] Except when the spacetime degenerates into a direct product of time and 3-space as in the Newtonian case. In general the metric of a 3-space t=const. is given by

$$\gamma_{\alpha\beta} = g_{\alpha\beta} + g_{0\alpha}g_{0\beta}/g_{00} \quad (\alpha, \beta = 1, 2, 3).$$

(In the General Theory this defines the metric of 'light geometry'.)

[13] From the mathematical point of view the two kinds of transformations refer to *dual spaces* related by the quantities

$$h_\alpha^a = \frac{\partial X^\alpha}{\partial x^\alpha} \tag{1}$$

and

$$h_a^\alpha = \frac{\partial x^\alpha}{\partial X^a}, \tag{2}$$

where the Greek indices refer to coordinate systems and the Latin indices to frames. To every *'space tensor'* (tensor under coordinate transformations) $T_{\varkappa\lambda\ldots}^{\alpha\beta\cdots}$ there exists a *'frame tensor'* (tensor under frame transformations) $T_{kl\ldots}^{ab\cdots}$ and vice versa according to

$$T_{kl\ldots}^{ab\cdots} = h_\alpha^a h_\beta^b \ldots h_k^\varkappa h_l^\lambda \ldots T_{\varkappa\lambda\ldots}^{\alpha\beta\cdots} \tag{3}$$

$$T_{\varkappa\lambda\ldots}^{\alpha\beta\cdots} = h_a^\alpha h_b^\beta \ldots h_\varkappa^k h_\lambda^l \ldots T_{kl\ldots}^{ab\cdots}. \tag{4}$$

In particular, the metrical tensors η_{ab} and $g_{\alpha\beta}$ defined by

$$(dS)^2 = \eta_{ab}\, dX^a\, dX^b = g_{\alpha\beta}(x)\, dx^\alpha\, dx^\beta \tag{5}$$

are related by

$$\eta_{ab} = h_a^\alpha h_b^\beta g_{\alpha\beta} (= \pm 1) \tag{6}$$

$$g_{\alpha\beta} = h_\alpha^a h_\beta^b \eta_{ab}. \tag{7}$$

Quantities like (5) that are invariant under both coordinate and frame transformations are to be called *proper invariants*. The frame tensors (3) have direct physical meaning since their values do not depend on the arbitrary choice of the coordinate system; they are the quantities that may be measured in a local frame.

The space tensors (4) have no direct physical meaning; but of course space tensor equations have, since $T_{kl\ldots}^{\alpha\beta\cdots} \equiv 0$ implies $T_{kl\ldots}^{ab\cdots} \equiv 0$.

For more comprehensive treatment cf. Lubkin (1961), Treder (1966), Pirani (1957).

[14] Our 'T' means frame time (extended time), whereas our 't' is a general time coordinate.

[15] The decision what interval is to be considered 'infinitesimal' depends on the curvature R at the given world point: the condition is $|dS|:R^{-2} \ll 1$. – The condition $|dS| \ll 1$ would have no meaning since $|dS|$ is a *dimensional* quantity (length). Thus, the existence of a second dimensional invariant R is necessary for a consistent and meaningful physical interpretation of the General Theory.

[16] These technical terms are misleading since they suggest motion between the frames.

[17] Any set γ may be said to be the *direct product* of sets γ_1 and γ_2 if and only if any element of γ is a pair of elements, one from γ_1 and one from γ_2, and vice versa.

[18] This group K_1 is given by the frame transformations

$$\Sigma^{(k)} \to \Sigma^{(i)} : \begin{cases} X^{(i)} = X^{(k)} + \mathbf{f}^{(ik)}(T) \\ T^{(i)} = T^{(k)} = T \end{cases}$$

with (group conditions)

$\mathbf{f}^{(ii)}(T) \equiv 0$ (identity)
$\mathbf{f}^{(ik)}(T) = - \mathbf{f}^{(ki)}(T)$ (inverse)
$\mathbf{f}^{(il)}(T) = \mathbf{f}^{(ik)}(T) + \mathbf{f}^{(kl)}(T)$ (composition).

If $\mathbf{f}^{(ik)}(T) = \mathbf{v}^{(ik)}T$, one obtains the familiar (proper) Galilei group. In general, $\mathbf{f}^{(ik)}(T)$ contains all time derivatives, as may be seen from serial expansion. Thus the (proper) Galilei group is but a small subgroup of K_1.

[19] The group K_2 is given by the frame transformation

$$\Sigma^{(k)} \to \Sigma^{(i)} : \begin{cases} X^{(i)\,a} = \omega^{(i)\,a}_{(k)\,b}(T)\, X^{(k)\,b} \\ T^{(i)} = T^{(k)} = T. \end{cases}$$

with (group conditions)

$\omega^{(i)\,a}_{(i)\,b} = \delta^a_b$ (identity)

$\omega^{(k)\,a}_{(i)\,b} \times \omega^{(i)\,b}_{(k)\,c} = \delta^a_c$ (inverse)

$\omega^{(i)\,a}_{(l)\,c} = \omega^{(i)\,a}_{(k)\,b} \times \omega^{(k)\,b}_{(l)\,c}$ (composition).

[20] A simply connected n-dimensional space of maximal symmetry admits $N = n(n+1)/2$ independent symmetry operations; this gives $N = 10$ for $n = 4$. But Newtonian space-time is not a simply connected 4-dimensional space. (An element of Newtonian space-time is not a point but a pair of points.)

[21] This was first shown by Cunningham (1910) and Bateman (1910). Cf. also Cashmore (1963) and Strauss (1966), p. 110.

[22] This allows for additional conservation laws (such as that for electric charge).

[23] Cf. Alexander (1956), Leibniz (1691–95).

[24] *Not* vindicated is Newton's conception of *'absolute space'*. However, this conception does not play any role in Newtonian mechanics: 'absolute acceleration' means acceler-ation with respect to any one of the inertial frames. For this reason it is usually held to be a metaphysical construct. Yet 'absolute space' as conceived by Newton is an *inconsistent concept*: on the one hand it is supposed to be homogeneous and isotropic like the 'relative spaces' of our experience, on the other hand a displacement in absolute space is supposed to correspond to a real or fictitious process, which implies inhomo-geneity. Thus, a mathematical model of Newton's 'absolute space' is impossible. To turn Newton's conception of 'absolute space' into a consistent notion it has to be re-interpreted to mean *'uniquely determined preferential frame'*. (The Maxwell equations were once thought to define such a frame.)

[25] Cf. Dautcourt (1964) and the literature quoted there.

[26] Mach (1883), quoted after Heller (1964), p. 47.

[27] Mach (1883), quoted after Heller (1964), p. 32.

[28] Schilpp (1949), p. 52.

[29] Einstein (1916b), quoted after Heller (1964), p. 156.

[30] Milne (1948).

[31] Strauss (1957/58); cf. also Strauss (1966).

[32] Mach (1883), quoted after Heller (1964), p. 34.

[33] Mach (1883), quoted after Heller (1964), p. 39.

[34] The exclusion of conservative forces depending on velocity by Newtonian mechanics was one of the reasons for Maxwell to reject the mechanical theory of Weber. He wrote: "(2) The mechanical difficulties, however, which are involved in the assumption of particle action at a distance with forces which depend on their velocities are such as to prevent me from considering this [Weber's] theory as an ultimate one. ...". (quoted after Magie (1935), p. 529).

[35] "Machsches Prinzip: Das G-Feld [$g_{\mu\nu}$-field, M.S.] ist *restlos* durch die Massen der Körper bestimmt. Da Masse und Energie nach den Ergebnissen der Speziellen Relativitätstheorie das gleiche sind und die Energie formal durch den symmetrischen Energietensor ($T_{\mu\nu}$) beschrieben wird, so besagt dies, daß das G-Feld durch den Energietensor der Materie *bedingt* und *bestimmt* sei". (Einstein, 1918).

[36] Hönl (1966a, b).

[37] Even in his 1918 paper, quoted in Ref. 35, Einstein calls his 'Mach Principle' a generalization of "die Machsche Forderung, daß die Trägheit auf eine Wechselwirkung der Körper zurückgeführt werden müsse".

[38] For a somewhat fuller discussion of this point cf. Strauss (1967a).

[39] For the genesis of Planck's quantum theory cf. Strauss (1960).

[40] Einstein (1916) thought that the postulate of general covariance "dem Raum und der Zeit den letzten Rest physikalischer Gegenständlichkeit nehmen würde" and that only 'coincidences', i.e., the points of intersection of world lines, would remain as objective facts.

CHAPTER XVI

THE LOGIC OF COMPLEMENTARITY
AND THE FOUNDATION OF QUANTUM THEORY*

INTRODUCTION

Several attempts have been made to provide an axiomatic basis for the statistical transformation theory in quantum physics in the form of simple general principles. Thus, in his well-known book on *Quantum Mechanics* Dirac uses the *superposition principle* as a fundamental principle. This principle permits indeed to determine many characteristic features of the mathematical formalism. It does not, however, suffice to determine even the algebra of the state calculus, as Dirac has noticed himself. From the present point of view the superposition principle may be looked upon as an ingenious but rather artificial formulation of complementarity.

A complete axiomatic foundation of the statistical transformation theory is due to VON NEUMANN [1]. The present work is closely related to it; its critical discussion is a natural starting-point.

Von Neumann's postulates demand essentially two things: (A) a one-one correlation between physical quantities and hypermaximal Hermitean operators in Hilbert space, and (B) linearity of the mean value operator for these quantities.

This deduction of the statistical formulae of quantum mechanics appears to be remarkable and satisfactory in so far as it makes no use of hypotheses concerning equal or numerical probabilities – in contrast to other statistical theories, in particular classical statistical mechanics. From the point of view of the *Correspondence Principle* this had to be expected since numerical probabilities would have no analogues, in the sense of that Principle, in classical mechanics.

However, from a *physical* point of view it can hardly be called satisfactory to base the theory on a postulate whose connection with experimental facts is as little intelligible as is the case with postulate (A). Instead, one would like to see a principle directly suggested by experience, as in Thermodynamics or Relativity Theory, such as the principle of indeterminacy or complementarity.

Even when adopting the *formal* point of view one is struck by the circumstance that the Principle of Complementarity, so closely connected[2] with that of Correspondence, is merely implicitly contained in postulate (A) but is not at all involved in the postulates concerning quantum mechanical probabilities. This would not be surprising if complementarity had no bearing on the mathematical theory of probability – a condition that is not satisfied. In fact, *complementarity restricts the validity or applicability of the ordinary theory of probability in a quite definite manner.* This may be seen even without the use of the formalism from the following examples.

Consider a statistical ensemble of hydrogen atoms all in the same energy state E_n. Then there exist the probabilities $\mathrm{prob}(E_n; I_{\Delta q})$ and $\mathrm{prob}(E_n; I_{\Delta p})$ to find the value of Q within the interval $I_{\Delta q} = (q, q + \Delta q)$ or the value of P within $I_{\Delta p} = (p, p + \Delta p)$, respectively. According to the ordinary theory of probability there would then also exist the probability $\mathrm{prob}(E_n; I_{\Delta q} \text{ and } I_{\Delta p})$ for finding both the value of Q within $I_{\Delta q}$ and the value of P within $I_{\Delta p}$. In view of the complete[3] complementarity between P and Q this probability could not be tested. The formalism yields for it a two-valued complex expression – it gives a nonsensical answer to a senseless question. This violation of the ordinary calculus of probability (or rather: its rules of existence) does not destroy the internal consistency of the calculus. This consistency would be destroyed only if either the probability $\mathrm{prob}(E_n \text{ and } I_{\Delta q}; I_{\Delta p})$ or $\mathrm{prob}(E_n \text{ and } I_{\Delta p}; I_{\Delta q})$ would exist (which is not the case, due to the complementarity between H and Q or H and P, respectively) because then $\mathrm{prob}(E_n; I_{\Delta q} \text{ and } I_{\Delta p})$ would be numerically determined by the general multiplication theorem:

$$\mathrm{prob}(I_{\Delta a}; I_{\Delta b} \text{ and } I_{\Delta c}) = \mathrm{prob}(I_{\Delta a}; I_{\Delta b}) \, \mathrm{prob}(I_{\Delta a} \text{ and } I_{\Delta b}; I_{\Delta c})$$
$$= \mathrm{prob}(I_{\Delta a}; I_{\Delta c}) \, \mathrm{prob}(I_{\Delta a} \text{ and } I_{\Delta c}; I_{\Delta b})$$

Cases where two (or all three) prob expressions with a logical conjunction exist do occur in quantum mechanics, e.g. in the case $A = M_x$, $B = Q_x$, $C = P_x$ (x-components of angular momentum, position and linear momentum vector, respectively). In this case, where A commutes with both B and C, the probabilities on the two right-hand sides exist while that on the left-hand side does not. [Hence the formalism should give a real one-valued expression for these probabilities, as indeed it does, but it can be interpreted only in the sense of the two right-hand sides.] (Of course, one may measure B in one half of an ensemble and C in the other half and multiply the relative frequencies corresponding to $\text{prob}(I_{Aa}; I_{Ab})$ and $\text{prob}(I_{Aa}; I_{Ac})$; but this is something quite different from a proper application of the general multiplication theorem.)

These simple examples demonstrating the limited applicability of the ordinary calculus of probability should make it clear that the *mean value postulate* [used by von Neumann] *is not an equivalent substitute for the rules of the calculus of probability*, and, hence, that it does not suffice to clarify the relation between that calculus and quantum mechanics which has been the original aim of von Neumann's work. (To be sure, this inequivalence has nothing to do with the question whether mean values are an equivalent substitute for a probability distribution: the latter is indeed determined by the mean values of all 'momenta' of the quantity in question.) The point is that the *logical operations* of the calculus of probability cannot be immitated by the averaging operation. The relation between the calculus of probability and the calculus of mean values is not one-one: only the former determines the latter.

Now it could happen that a physical theory permits only mean values to be measured. In that case a trunkated theory of probability, characterized by the postulate of the linearity of the mean value operation, would suffice. Contrary to a previous stage in the physical interpretation, quantum mechanics is not such a theory. True, the ordinary calculus of probability demands too much, but the mean value postulate [of von Neumann] demands too little; the ordinary calculus of probability is still needed if all quantities concerned commute.

It may seem paradoxical in view of all this that the mean value postulate suffices for deducing the correct formulae. In this connection it must be pointed out that in von Neumann's deduction essential use is made of an *extension* of the mean value postulate to 'quantities' that are represented by projection operators; as shown by von Neumann, these projection operators represent *statements* on the measured values of these quantities rather than the quantities themselves – (to the eigenvalues 1 and 0 of the projection operators correspond the truth values 'true' and 'false', respectively); thus, the calculus of the projection operators represents a kind of sentential calculus. Now it emerges from recent investigations concerning the axiomatics of probability theory that the algebra of the ordinary sentential calculus can be used as a substitute for certain axioms in the theory of probability; thus, given that all probabilities concerned exist, addition and multiplication theorem may be deduced from one another by employing the distributive laws of the sentential calculus. [[...]]. Although the state of affairs is somewhat different when we turn to quantum mechanical measurement statements and the projection operators representing them – the *logical meaning of complementarity* resides in just this difference – the difference, when properly formulated, does *not* concern the *algebraic formulae* as such but merely their *range of applicability*, viz., questions of *existence*. In this way it becomes intelligible that the mean value postulate, extended to projection operators, does suffice for deducing the statistical formulae and that, on the other hand, the anomalies mentioned above do remain. These anomalies present a violation of the axioms of probability theory only if the existential axioms corresponding to the ordinary sentential calculus are included (REICHENBACH[4]) or – what amounts to the same – if the region of definition of the probability function is supposed to be a set system [viz., the set of all subsets of a given set] (KOLMOGOROFF[5]). In other words: *the restricted applicability of the ordinary theory of probability is due solely to the invalidity of the ordinary sentential calculus for quantum mechanical measurement statements.* [[...]]

Since for clarifying the relation between probability theory and quantum mechanics it suffices to heed complementarity, the combination of complementarity and probability theory may be expected to be sufficient for building up the general formalism of the statistical transformation theory so that von Neumann's postulate (A) may be replaced by the Principle of Complementarity.

How far this expectation is justified will emerge from the following.

I. THE LOGICAL FORMULATION OF COMPLEMENTARITY

(1) If complementarity is to be used for an axiomatic reconstruction of quantum theory it has to be formulated in a way suitable for formal operation. As long as complementarity is conceived primarily as a relation between physical quantities it is difficult to see how this should be done; there is no obvious reason why quantities that cannot be measured

simultaneously should be represented by operators in Hilbert space.

The statement that two quantities cannot be measured simultaneously may be expressed thus: two statements concerning the results of measurement of the two quantities cannot be decided both, or: deciding the one makes it impossible to decide the other, or: decidability of one implies undecidability of the other. In this way complementarity becomes primarily a [semantic] *relation between statements*. This makes it possible to formulate complementarity in a formal way [viz., to formulate a non-classical sentential or predicate calculus to be called *complementarity logic*].

(2) What we need is not an axiomatic system for the sentential calculus but its algebra which can easily be recognized as rules of ordinary language; the semantic definition of complementarity as given under (i) then leads to a modified sentential calculus, this modification being the formal [syntactic] expression of complementarity.

As sentential variables we use the letters

$$R, S, T, \ldots$$

and for the negation and the sentential connectives we use Russell's symbols:

(a) $\sim$ for 'not' (negation)
(b) $\cdot$ for 'and' (conjunction, logical product)
(c) $\vee$ for 'or' (disjunction, logical sum)
(d) $\equiv$ for 'if and only if' (equivalence)

but, following Hilbert, we shall put the negation sign above the sentential symbol.

We then have the following equivalences [characteristic of Boolean algebra]:

$$\text{n1.} \quad \tilde{\tilde{R}} \equiv R$$

(1)	$R \cdot R \equiv R$	
(2)	$R \vee R \equiv R$	
(k1)	$R \cdot S \equiv S \cdot R$	L
(k2)	$R \vee S \equiv S \vee R$	

$$(a1) \qquad R \cdot (S \cdot T) \equiv (R \cdot S) \cdot T$$
$$(a2) \qquad R \vee (S \vee T) \equiv (R \vee S) \vee T$$
$$(d1) \qquad R \vee (S \cdot T) \equiv (R \vee S) \cdot (R \vee T)$$
$$(d2) \qquad R \cdot (S \vee T) \equiv (R \cdot S) \vee (R \cdot T).$$

$\left.\right\} \textbf{L}$

The algebraic significance of the negation consists in the fact that it permits to solve the equivalences

$$X \vee (R \cdot S) \equiv R \qquad\qquad Y \cdot (R \vee S) \equiv R$$

for X and Y:

$$X \equiv R \cdot \tilde{S} \qquad\qquad Y \equiv R \vee \tilde{S}.$$

Thereby the expressions

$$O =_{df} R \cdot \tilde{R} \qquad\qquad E =_{df} R \vee \tilde{R}$$

play the role of zero and unity:

$$(n1.1) \quad S \vee O \equiv S \qquad\qquad (n2.1) \quad S \cdot E \ \equiv S$$
$$(n1.2) \quad S \cdot O \ \equiv O \qquad\qquad (n2.2) \quad S \vee E \equiv E.$$

O is called *contradition* and E *tautology*. [[. . . . .]]

The equivalences **L** can be handled in the same way as algebraic equations, which implies the following rule of substitution for the variables $R, S, T, \ldots$:

L Subst. 'R' may be replaced by (a) any other sentential symbol such 'S', (b) '$\tilde{R}$', (c) '$S \cdot T$', (d) '$S \vee T$', (e) '$S \equiv T$', [any such substitution for 'R' to take place everywhere where 'R' occurs within a given formula].

The calculus defined by the equivalences **L** and the rule **L Subst.** will be called **L**-*calculus*.

(3) Now the **L**-calculus is just that part of the ordinary sentential calculus that can be maintained if complementarity is taken into account, with the following proviso. According to the semantic definition of complementarity as given under (1) the sentential connection of two complementary sentences gives an undecidable statement and hence a meaningless sentence, contrary to what is implied in the ordinary sentential calculus. Hence the equivalences **L** must be interpreted thus: if one (and hence also the other) side of an equivalence is meaningful the equivalence is logically true; if this condition is not fulfilled the equivalence is not false but meaningless.

We now [take the *decisive step* and] *forbid* the formation of meaning-less expressions. The resulting calculus will be called **L'**. Then the situation is as follows:

Though the equivalences **L** can be maintained [they are not violated semantically], the domain of definition of the sentential connectives is no longer a 'field'. Through this *change in algebraic structure* the **L'**-calculus is *not isomorphic with the set calculus*: while section and junction of two given sets always exist the corresponding conjunction and disjunction of two sentences may not exist in the **L'**-calculus. [[. . . . .]]

(4) The semantic justification for ruling out the formation of a compound sentence as given above breaks down if the two measurement statements refer to different instances of time: since measurements of complementary quantities can be performed at different instances of time, the corresponding statements can be decided both. Let us call such statements *simply complementary* to each other in contradistinction to two complementary statements referring to the same instant of time which will be called *strictly complementary*. Is there any reason for ruling out the sentential connection of simply complementary statements? The answer to this question is bound up with the following consideration.

The statement of a probability relation between strictly complementary statements has no obvious or direct meaning; it cannot be decided because complementary quantities cannot be measured simultaneously. An [experimental] meaning can be attached to it only by a limiting process $t_2 \rightarrow t_1$ when one of the two quantities is measured at t_1 and the other at t_2. Hence, if compounds of simply complementary statements were admitted [we would have a logical discontinuity for $t_2 \rightarrow t_1$, and] extending the non-admittance of compounds from strictly to simply complementary sentences could be justified only [by the wish to remove this discontinuity or] by reference to the calculus of probability, namely by the obvious demand that the prob expressions be continuous functions of time. No independent logistic meaning would then attach to simple complementarity.

Now, although the compound 'position of S at t_1 is within $I_{\Delta q}$ and momentum of S at t_2 is within $I_{\Delta p}$' is meaningful in sofar as it can be decided experimentally, it will not occur in a rationally constructed language [of quantum mechanics], because *the consequences* [predictions] *to be drawn from either part separately contradict each other even when the two parts are true.* Hence, only one *or* the other part can occur in any correct deduction. This is the formal expression of what is called *Nichtobjektivierbarkeit* of measurement results.

(The paradox resulting from handling complementary statements according to the rules of ordinary logic have lately been exposed by EINSTEIN[6] and SCHROEDINGER[7].) [[. . . .]]

(5) The sentence 'The momentum of the particle lies within $I_{\Delta p}$' does not characterize an individual state of affairs but a class [of particles]. Hence our $R, S, T, \ldots$ are to be regarded as *class* [or *predicate*] variables. This does not interfere with the equivalences and the substitution rule: the two calculi are isomorphic. The difference between the **L**- and the **L**'-calculus is of course transfered to the class [or predicate] calculus so that we have to distinguish between the ordinary class [or predicate] calculus corresponding to **L** and the one corresponding to **L**'; the latter may be called complementary class [or predicate] calculus. [[...]]

II. CALCULUS OF PROBABILITY

(6) We now turn to the calculus of probability. We show first why the (essentially equivalent) axiomatic systems of Reichenbach and Kolmogoroff cannot be used when complementarity is taken into account. According to Reichenbach (l.c.) a probability statement is a general implication between sentences stating class membership of elements and hence written in the form

$$\textbf{R} \qquad (i)\,(x_i \in 0 \underset{p}{\Longrightarrow} y_i \in P)\,; \tag{R1}$$

here, O and P are class variables, x and y are individual variables, $\Rightarrow$ is the sign for the prob relation ('probability implication'), and p is the numerical value of the probability. A short-hand designation for (R1) is

$$\textbf{R} \qquad (0 \underset{p}{\Longrightarrow} P) \quad \text{or} \quad w\,(0, P) = p. \tag{R2}$$

The multiplication theorem (axiom IV) may then be written in the form

$$\textbf{R} \qquad (0 \underset{p}{\Longrightarrow} P)\cdot(0\cdot P \underset{u}{\Longrightarrow} Q) \supset (0 \underset{\underline{w}}{\Longrightarrow} P\cdot Q)\cdot(w = pu). \tag{R IV}$$

Here, a compound of P and Q occurs only on the right-hand side of the implication, and this (together with the rule of inference referred to above) makes it possible to deduce from meaningful expressions an expression that may be meaningless; in other words: **R** IV has the property of transfering existence.

Similarly, the first axiom of Kolmogoroff's system demands that the domain of definition of the prob function be a field of sets, i.e., that with any two sets also their junction (set sum), their difference and their section belong to it, thus, the algebra of the ordinary sentential [or predicate] calculus is presupposed here, too. Hence this axiomatic system cannot be used either if complementarity is taken into account.

(7) For complementarity logic we have used a system of equivalences from ordinary logic; similarly we must use a system of equations from ordinary probability theory as basis for complementary probability theory. Such a system has already been given by Reichenbach (l.c.); it reads:

$$\mathbf{W} \begin{cases} \text{I.1} \;\; w(R, R \vee S) = 1 \\ \text{I.2} \;\; w(R, S \cdot \tilde{S}) = 0 \\ \text{I.3} \;\; 0 \leq w(R, S) \\ \text{II.} \;\;\; w(R, S \vee T) = w(R, S) + w(R, T) - w(R, S \cdot T) \\ \text{III.} \;\;\; w(R, S \cdot T) = w(R, S)\, w(R \cdot S, T) \end{cases}$$

Though here (III), too, one side of the equation may be meaningless while the other one is not, this does no harm since the equation sign – contrary to the implication – does not transfer existence; it merely implies that certain probabilities are equal if they exist. Hence meaningless prob statements cannot be deduced from meaningful ones.

In order to compensate for the loss of deductive power in the transition from the **R**-system to the **W**-system one has to postulate in the ordinary prob calculus that a probability exists if its numerical value is determined by the equations of the calculus [and other probabilities known or assumed to exist] (Reichenbach's rule of existence, Kolmogoroff's 1. axiom). Similarly, we need the following *Existential Postulate*: if the numerical value of a prob function $w(R; S)$ is determined according to the **W**-system by other probabilities known to exist, then $w(R; S)$ exists *provided that both R and S exist*. [[...]]

Let us look for a moment at the [classical] system **LW** involving the ordinary sentential [or rather predicate] calculus. This system, as is well known, admits a set theoretical interpretation as follows:

To the tautology E corresponds a basic set $\underline{E}$, to the contradiction O the empty set $\underline{O}$, to the sentences [or rather predicates] $R, S, \ldots$ correspond subsets $\underline{R}, \underline{S}, \ldots$ of $\underline{E}$, to the negation $\tilde{R}$ corresponds the complementary set $\underline{E} - \underline{R}$, to the conjunction $R \cdot S$ corresponds the set section $\underline{R} \cdot \underline{S}$, and to the disjunction $R \vee S$ the union $\underline{R} \dotplus \underline{S} = \underline{R} + \underline{S} - \underline{R} : \underline{S}$. Every additive set function

$$P(\underline{R} + \underline{S}) = P(\underline{R}) + P(\underline{S}) \tag{P1}$$

with

$$w(\underline{R}, \underline{S}) = P(\underline{R} \cdot \underline{S})/P(\underline{R}) \tag{P2}$$

then satisfies the axioms **W**, so that the system **W** may be replaced by (P1), (P2) together with suitable existential postulates (Kolmogoroff, loc. cit.).

III. QUANTUM THEORY

(8) We are now going to characterize the domain of definition of the [quantum mechanical] prob function. As can be seen directly from experimental experience, we are confronted with the following facts:

Q (a) To every measurement propostition R there exist an infinite number of other measurement propostitions, all noncomplementary to one another and to R (e.g., all those resulting from R by replacing the measurement interval refered to in R by a *larger* one).

Q (b) To every measurement proposition R there exist an infinite number of measurement propositions S_i all complementary to R (e.g., all those resulting from one such S by replacing the measurement interval referred to in S by a smaller one.

Q (c) The relation of [sentential or predicational] connectibility (σ) and the relation of inconnectibility (κ) are neither transitive nor intransitive; (i.e., all four possibilities of the scheme (see top of next page) are realized in nature, e.g., by the examples given in the last three columns; Q_x, P_x, M_x are components of position, momentum, and angular momentum, respectively, and for the pertaining intervals any finite intervals may be choosen.)

Taken together $Q(a)$–$Q(c)$ imply that our $R, S, T, \ldots$, form an in-

RS	ST	RT	R	S	T
σ	σ	σ	Q_x	Q_y	Q_z
σ	σ	κ	Q_x	M_x	P_x
κ	κ	σ	M_x	M_y	Q_x
κ	κ	κ	M_x	M_y	M_z

finite (in fact: continuous) domain, with infinite 'islands' in which the ordinary sentential [or rather predicate] calculus and hence the unrestricted calculus of probability holds. [The algebraic structure of the domain is thus that of a *partial Boolean algebra*].

As the cardinality of the domain is that of a continuum, the following postulate appears adequate:

Q *con.* The prob function w is a continuous function of time [or rather the time interval(s) occurring] and of the measurement intervals.

(9) We are now going to show: *the calculus of projection operators over a linear vector space is isomorphic to the* **L**$'$*-calculus* under the following mapping:

$$\mathbf{Z} \begin{cases} & \text{predicates} & \text{projection operators} \\ 1. & R & R \\ 2. & \tilde{R} & I - R \\ 3. & R \cdot S & RS \\ 4. & R \vee S & R + S - RS \end{cases}$$

(I is the identity operator satisfying $IR = R$ for all R).

(a) The logical equivalences **L** turn into mathematical identities if the predicates are replaced by projection operators according to **Z**; note that RS and $R + S - RS$ are projectors only if $RS = RS$.

(b) If $RS \neq SR$, RS is not a projector, $(RS)(RS) \neq RS$, and hence the predicational compounds formed with R and S are not predicates either, i.e., R and S are complementary to one another. If this is taken into account the mapping **Z** is one-one.

[Confusion may arise from the fact that there is a one-one relation between projectors and closed linear subsets of the linear vector space concerned: this may suggest to take the calculus of closed linear subsets – instead of the calculus of projectors – as the mathematical model of quantum logic. The following paragraph shows why this is not feasible and thus refutes the much discussed 'Logic of Quantum Mechanics' of Birkhoff-von Neumann which likewise appeared in 1936.]

Since there is a one-one mapping between projectors and the [closed] linear subsets of the vector space $\mathfrak{R}$ concerned, **Z** implies a one-one relation between the predicates $R, S, \ldots$ of complementarity logic and the closed linear subsets of $\mathfrak{R}$. This corresponds to the isomorphism between the ordinary sentential calculus and the ordinary set calculus: instead of arbitrary sets we have now closed linear subsets of a vector space. However, this analogy is rather limited: the calculus of the closed linear subsets is *not isomorphic* to [the calculus of projectors and to] the calculus **L′**. True, the *junction* of two linear subsets is again a linear subset if and only if the pertaining projectors commute, but the *section* of two linear subsets is always a linear subset, even when the pertaining projectors do *not* commute, i.e., even when the predicates concerned are complementary so that the compound predicate does *not* exist.

Thus, it is *not* possible to satisfy the system **L′** by linear subsets if isomorphism is to be maintained. This is decisive for the following treatment: the prob function w cannot be [represented by] a set function.

(10) By virtue of **Z** the domain of definition of the prob function w may be taken to be the set of projection operators [or rather the direct product of this set with itself]. This however does not imply that the numbers $w(R, S)$ can be determined otherwise than by explicit coordination [i.e., on a purely empirical basis]. To obtain a [physico-] mathematical theory we must demand that there exist a general function $W(R, S)$ depending only on R and S, which satisfies the system **W** with

$$w(R, S) = W(R, S);$$

in other words: *the equations* **W** *are to be considered as functional equations for W over the set of projection operators*. As the values of W are to be real numbers, this implies that the vector space concerned is a metrical space. Though the metric is not uniquely determined by **W** alone a simple postulate to be given later will fix it.

We solve the functional equation **WIV** by

$$W(R, S) = \frac{f(RS)}{f(R)} \tag{W 1}$$

which corresponds to (P2).

Substituting this in **W** III gives, in consideration of **Z**,

$$f(RS + RT - RST) = f(RS) + f(RT) - f(RST)$$

which yields the functional equation

$$f(R + S) = f(R) + f(S). \qquad \text{(W 2)}$$

Its [general] solution is

$$f(R) = c \, Tr \, R \quad (c = \text{constant independent of } R)$$

since the trace $Tr \, R$ is the only linear invariant that depends only on R. Hence

$$\boxed{W(R, S) = \frac{Tr \, RS}{Tr \, R}} \,. \qquad \text{(W 4)}$$

(W4) is the general expression for the quantum mechanical probabilities. It merely remains to fix metric and number of dimensions of the underlying vector space $\mathfrak{R}$.

(11) It is obvious that the metric of $\mathfrak{R}$ must be either Euclidean [i.e., real][9] or unitary; otherwise the trace $Tr \, R$, defined by $R_i^i = R_{ik} g^{ik}$, would depend on the metrical tensor g^{ik} for which there would be no physical interpretation.

For deciding between Euclidean [real] and unitary metric we consider the expression $Tr \, RST$ which occurs in the general multiplication theorem. In the case of *Euclidean* [real] metric this expression is always real-valued, even if none of the projectors R, S, T commutes with any of the other two, i.e., if none of the three expressions of the multiplication theorem have any physical meaning. In the case of *unitary* metric the said expression is complex-valued (and the complex-conjugate of $Tr \, RTS$) iff none of the three projectors commutes with any of the other two. Thus, *only the choice of unitary metric is in accoord with complementarity logic.*

[This result is of fundamental importance in two respects. For one, it answers the question, first put to the author by Reichenbach, whether the use of complex-valued state functions in quantum mechanics is a mathematical trick that could be avoided in principle (as often in classical physics), and if not, why not. Second, it shows why all attempts at interpreting the quantum mechanical formalism in terms of classical probability or statistics are doomed to failure.] [[...]]

(12) Finally, the number of dimensions [of the linear vector space $\mathfrak{R}$] can be determined by the well-known commutation rules for canonical quantities or else by the postulate that there exist continuous regions of measurable values; either postulate leads to an infinite number of dimensions. To the latter postulate correspond our axioms $\mathbf{Q}(a, b)$. It is easy to show that they demand an infinite number of dimensions. (Incidentally, it would be difficult to attach any physical meaning to a finite number of dimensions). Thus, besides (W4) we have also established the Hilbert space. [[...]]

(13) In conclusion it should be pointed out that nothing has been said about the connection between the projectors [which were introduced as a mathematical model of complementarity logic] on the one hand and the [hypermaximal Hermitean operators representing] physical quantities on the other hand: the question which projector corresponds to a given measurement statement [or predicate] has been left open. Quite naturally, this question can only be decided by considerations of correspondence. [It should be noted, however, that from our point of view the projectors are more fundamental than the hypermaximal Hermitean operators. This is in line with the fact that the later can be defined in terms of the former – a fact that would be merely a mathematical curiosity if the projectors had no fundamental significance.]

The considerations given above confirm and substantiate the often stressed analogy between the theory of special relativity and quantum mechanics: just as the world geometry of Einstein-Minkowski merely expresses the existence of a finite upper limit c of signal velocities, thus the general formalism of quantum mechanics merely reflects the unavoidability of complementarity resulting from the existence of the finite quantum of action h. This formalism thus appears as the appropriate mathematical language for expressing all special quantum mechanical experience. From this point of view it is rather obvious that the formalism has stood the test of the many-body problem and of relativistic generalization, and the same point of view may help to decide the question whether this general formalism is wide enough to encompass a future theory of elementary particles.

POSTSCRIPT 1971

The introduction of a non-classical logic in physics raises a number of philosophical questions, and the introduction of two competing logics for the same physical theory raises some additional questions of a more technical nature, but not without philosophical import. In the following notes I shall try to answer some of these questions.

(1) In the first place let me point out that there is no such thing as '*the* logic *of* quantum mechanics'. A physical theory is not given in the form of a formalized language but as the union of a mathematical formalism and its physical interpretation; the formalism does not contain any descriptive predicates (sentential functions) and hence no predicate connectives either. The connectives become part of a formal system only if the language of the theory is formalized. It follows that the logical syntax of a physical theory depends on the way the language of the theory is beeing formalized. Vice versa, advocating a particular 'logic' (viz. logical syntax) *for* a physical theory means advocating a particular way of formalizing its language. If different 'logics' are advocated for the same physical theory, it is only by comparing all consequences of the implied formalizations that a proper judgement on their relative merits can be given. True, even when we have a complete list of all relevant differences we may not agree on their relative merits but at least we are then compelled to state our reasons for any preferential decision we care to make.

(2) With this in view, I have carried out the two formalizations corresponding to complementarity logic (partial Boolean algebra) and 'quantum logic' (Birkhoff-von Neumann's nondistributive lattice algebra), respectively, in 1937 (doctor thesis, Prague 1939). Though all copies of this have been lost, one of its main results is easily established: *the Birkhoff-von Neumann logic leads to a language containing 'metaphysical' sentences*, namely the conjunction of sentences that are inconnectible in complementarity logic.[10]

(3) There are other – and perhaps more important – reasons for preferring complementarity logic to nondistributive lattice logic. Here are some of them: –

(a) *Giving up the distributive law for the sentential connectives implies giving up the (semantical) two-valuedness*: in any two-valued logic the two sides of the distributive law have the same truth-value. Now I have no philosophical objections against multi-valued logics, but none of the advocates of the Birkhoff-von Neumann logic seems to have noticed this implication.

(b) If the physical predicates are to be represented by subspaces (closed linear subsets) rather than by the projectors on these subspaces, one should expect that the quantum mechanical probabilities are functions of these subsets. However, they are functions of the projectors (and hence merely functionals of the subspaces).

(c) Most important of all, the Birkhoff-von Neumann logic *does not lead to the unitary metric*, even when combined with the prob calculus: it is equally well compatible with real (Euclidean) metric. On the other hand, complementarity logic demands unitary metric (complex valued state vectors), as shown above. As the unitary metric is one of the most important characteristics of the quantum mechanical state space, the Birkhoff-von Neumann logic, whatever it may be, is certainly not characteristic of quantum mechanics.

(4) In view of all this the question arises *why* the Birkhoff-von Neumann logic has attracted far more attention than complementarity logic. My answer: this is not just a case of authority against non-authority – after all, complementarity logic goes also back to von Neumann, if only by implication – but rather a case of fashion against unfashion. Indeed, lattice theory became quite fashionable in the nineteenthirties, thanks mainly to the work of Birkhoff, while partial Boolean algebra, of which the algebra of projectors and complementarity logic are examples, had to wait for another 30 years to become respectable among mathematicians. The quantum physicists, of course, have used complementarity logic all the time, even when not knowing it, and have paid no attention to 'the' 'quantum logic' of Birkhoff-von Neumann.

(5) Let me just add that complementarity logic has been rediscovered in recent years by several authors, among them P. SUPPES[11] and F. W. KAMBER[12].

(6) Does the use of a nonclassical logic in physics imply that logic is empirical, at least in the sense in which physical geometry is empirical, as argued, e.g., by H. PUTNAM[13]? My answer is 'no', as follows from what I have said above. The analogy with physical geometry breaks down because geometry belongs to the mathematical substructure of a

physical theory while logic does not; the question of 'logic' (viz., logical syntax) only arises in connection with an (implied or intended) formalization of physical language. Thus, logic is neither empirical, nor *a priori*, nor a matter of convention. Rather, it is a matter of optimal choice among a limited number of possibilities. It is only the whole set of possibilities that has some empirical content or significance. Of course, if we prescribe form and meaning of the (atomic) sentences the logic of the sentential connectives may only depend on the (semantic) meaning of the latter. But this is really a question that would require a separate paper, if not a monograph, for full treatment.

NOTES

*Translated from 'Zur Begründung der Statistischen Transformationstheorie der Quantenphysik', *Sitz. Ber. Berl. Akad. Wiss., Phys.-Math. Kl.* **27** (1936), 90–113.

[1] J. von Neumann, 'Wahrscheinlichkeitstheoretischer Aufbau der Quantenmechanik', *Goett. Nachr.* (1928), 245; *Mathematische Grundlagen der Quantenmechanik*, Berlin 1932, Kap. IV. See also M. Born und P. Jordan, *Elementare Quantenmechanik*, Berlin 1930, 6. Kap.

[2] [The true nature of this connection has only emerged much later in the study of the intertheory relations between quantum mechanics and classical Hamiltonian mechanics. The upshot is this: the Principle of Correspondence allows us to consider as meaningful statements of the form 'The value of a quantity Q of a physical system S lies within the interval (q_1, q_2)'; statements of this form referring to complementary quantities are then to be treated as complementary in the sense of complementarity logic, i.e., inconnectible, as shown in this paper. On the other hand, we may use instead statements of the form 'The physical system S is in a state where the quantity Q has a value between q_1 and q_2'; in this case statements referring to complementary quantities can be treated as contradictory (their conjunction would be allowed as meaningful but untrue) and the need for complementarity logic – or any other 'logic of quantum theory' – does not arise. The essential difference between the two statements emerges when their negations are considered: the negation of the first form would read 'The value of quantity Q of S lies outside the interval (q_1, q_2)' – this is indeed the proper negation for classical and the (improper) negation for quantum mechanics – but this is *not* equivalent to the (proper) negation of the second form which includes all states where the quantity Q has no value in any finite interval. The notion 'proper negation' as used here is a semantical one. A semantically adequate syntactic characterization of 'proper negation' for arbitrary languages has been attempted (e.g. by Carnap, *Logical Syntax of Language*, London 1937), but most attempts can be shown to be inadequate. In the present paper measurement statements are supposed to have the first form mentioned above.]

[3] [[...]] [Two quantities A and B are called *totally complementary* iff there is no state for which A and B have values within any finite intervals.]

[4] H. Reichenbach, 'Axiomatik der Wahrscheinlichkeitsrechnung', *Math. Z.* **34** (1932), 568; *Wahrscheinlichkeitslehre*, Leiden 1935.

[5] A. Kolmogoroff, 'Grundbegriffe der Wahrscheinlichkeitsrechnung', *Erg. Math.* II/3 (1933).

[6] A. Einstein, B. Podolsky, and N. Rosen, *Phys. Rev.* **47** (1935), 777.

[7] E. Schroedinger, *Naturwiss.* **23** (1935), 807, 823, 844.

[8] [If we take the *span* instead of the (set theoretical) sum we have the same situation as with the section: the span of two linear subsets is always again a linear subset (subspace). The set of subspaces then forms an orthocomplemented lattice in which – contrary to the Boolean lattice – the distributive laws do not hold. This is the 'Quantum Logic' advocated by Birkhoff and von Neumann. The negation in this logic has the same meaning as in our complementarity logic, i.e., it too is a nonproper negation, referring as it does to the orthogonal complement.]

[9] [To be sure, a Euclidean space is a point space, not a vector space. But we can define in it a vector space by distinguishing a fixed point as origin, i.e., by abandoning homogeneity (group of translations or displacements) while maintaining isotropy (group of rotations).]

[10] To be sure, the metaphysical nature of these compound sentences arises solely from the logical interpretation of the nondistributive lattice connectives as sentential (or predicate) connectives, and not by the admission of predicates correlated to the lattice elements (subspaces). Indeed, if we consider two subspaces X and Y with the (non-commuting) projectors P_X and P_Y, we have of course the further projector $P_{X \odot Y}$; but while $X \odot Y$ is interpreted by Birkhoff-von Neumann as conjunction of two predicates, $P_{X \odot Y}$ *cannot* be so interpreted because $P_{X \odot Y} \neq P_X \cdot P_Y$. Thus, complementarity logic does not omit any physically meaningful predicates (as is sometimes suggested) but prevents their misinterpretation.

[11] P. Suppes, 'Probability Concepts in Quantum Mechanics', *Phil. of Sc.* **28** (1961), 378–389; 'Logics Appropriate to Empirical Theories', in *The Theory of Models* (ed. by J. W. Addison, L. Henkin and A. Tarski), Amsterdam 1965, p. 364–375; 'Une logique non-classique de la méchanique quantique', *Synthese* **10** (1966), 74–85.

[12] F. Kamber, 'Die Struktur des Aussagenkalkuels in einer physikalischen Theorie', *Nachr. Akad. Wiss. Goettingen* **10** (1964), 103–124.

[13] H. Putnam, 'Is Logic Empirical?' in *Boston Studies in the Philosophy of Science*, vol. **V** (ed. by R. S. Cohen and M. W. Wartofsky), Dordrecht 1969, pp. 216–241.

CHAPTER XVII

THE PARADOXES OF QUANTUM PHYSICS
AND THE COMPLEMENTARY MODE
OF DESCRIPTION*

ABSTRACT. It is shown that, contrary to a widely held opinion, the complementary mode of description suggested by N. Bohr, cannot be reduced to, or justified by, Heisenberg's Principle of Indeterminacy. It is the consideration of *three*, not of *two*, measurements, not necessarily simultaneous, which compels us to adopt the complementary mode of description.

INTRODUCTION

It is generally assumed that the question of the logical foundation of quantum theory has been finally settled by the use of a certain non-classical form of language suggested by N. Bohr and called by him the complementary mode of description. If the latter is rightly understood this view is correct. However, neither has the syntactical nature [1] of the complementary mode of description itself been generally realized, nor have the underlying facts always been stated correctly. This emerges most clearly from the current treatment of the so-called paradoxes of quantum physics. These paradoxes can be divided into two classes: those in which only two, and those in which three (or more) physical quantities are taken into consideration. We shall see that only the paradoxes of the second class compel us to adopt the complementary mode of description whereas those of the first class admit of a rather trivial solution within the framework of classical syntax. Since Heisenberg's Principle of Indeterminacy is concerned with only two physical quantities it follows that the complementary mode of description cannot be reduced to, or justified by, that principle.

A re-consideration of the paradoxes of quantum physics has won new interest in view of J. VON NEUMANN's [2] recent proposal of an alternative form of language for the use of quantum theory. This alternative form of language, though likewise non-classical, avoids the complementary mode of description. It is true that this language offers a solution for the paradoxes of the first class. It will be seen, however, that those of the second class would remain unsolved.

Our results are not only in agreement with the mathematical theory but can most easily be obtained by a simple analysis of the latter. However, it seems to be more instructive to discuss the paradoxes first on their own merit. In this way a deeper understanding of the mathematical theory itself will be attained.

I. THE PARADOXES OF QUANTUM PHYSICS

1. *An instance of the First Class*

Let us consider a system s_0 consisting of two particles s_1 and s_2 between which an attraction is taking place; this means that for any value r of the distance R between s_1 and s_2 a potential energy $E_{pot}(r)$ exists so that

$$E_{pot}(r) < E_{pot}(r') \quad \text{if, and only if,} \quad r < r'. \tag{1.1}$$

If we take for s_1 and s_2 a proton and an electron, respectively, and hence for $E_{pot}(r)$ the function

$$E_{pot}(r) = - e^2/r \tag{1.H}$$

we get the model of a hydrogen atom. The following considerations, however, are independent of this specification.

Suppose s_0 to be in a state where the total energy W has a definite value w_i:

$$W = E_{pot} + E_{kin} = w_i, \tag{1.2}$$

E_{kin} being the total kinetic energy (with regard to the centre of mass, say, or, if it be prefered, to a fixed system of reference).

Now let us ask what can be said about the value of R if s_0 is in the state given by (2).

On account of the possible interchange between kinetic and potential energy, R, of course, cannot be expected to have a constant value. Yet whatever the mechanism governing this interchange may be, E_{kin} cannot be negative:

$$E_{kin} \geqslant 0, \tag{1.3}$$

and hence the value of R cannot exceed a certain limiting value, $r_{i\,max}$, which is given by $E_{kin} = 0$, or the equivalent condition $E_{pot} = w_i$:

$$r \leqslant r_{i\,max}, \quad E_{pot}(r_{i\,max}) = w_i. \tag{1.4}$$

Indeed, (4) follows directly from (1)–(3).

A particular consequence of (4) is that the probability is zero that s_0, when in a state given by (2), has an R greater than $r_{i\ max}$:

$$\text{prob}\,(w_i; r) = 0 \quad \text{for any} \quad r > r_{i\ max}.$$

It is this consequence which seems to be contradicted by quantum theory. As is well known the latter gives a finite probability for *any* value of r.

Thus it looks as if quantum theory contains a contradiction.

Similar paradoxes can be obtained in many other cases.

2. *Discussion of the Previous Paradox*

We found that in classical theory the two statements

> 's_0 is at t_0 in a state where the energy W has a definite value w_i'   (S₁)

and

> 's_0 is at t_0 in a state where the distance R is greater than $r_{i\ max}$'   (S₂)

are incompatible with one another. In consequence of this we have

$$\text{prob}\,(w_i; r) = 0 \quad \text{for any} \quad r > r_{i\ max}. \qquad (2.1)$$

Quantum theory, on the other hand, seems to give

$$\text{prob}\,(w_i; r) \neq 0 \quad \text{for some} \quad r > r_{i\ max}. \qquad (2.2)$$

(2.1) and (2.2) contradict each other, and hence only one of them can actually belong to quantum theory.

Now it is well known that the probabilities predicted by quantum theory concern the values of physical quantities *found by actual measurement*. Without subscribing to any particular 'theory of measurement' we must admit that the measuring arrangement, when actually applied to the system s_0, constitutes an additional condition which has not been accounted for in (2.2). Hence (2.2) must be replaced by a statement of the form

$$\text{prob}\,(w_i, M^R; r) \neq 0 \quad \text{for some} \quad r > r_{i\ max}, \qquad (2.2')$$

where 'M^R' represents the condition that a measurement of R is being made. Obviously, (2.2') does not contradict (2.1).

Since the argument leading to (2.1) was on classical lines it may still be doubted whether (2.1) holds in quantum theory. As (2.1) is a logical consequence of the incompatibility of S_1 and S_2 we might as well ask whether S_1 and S_2 are still incompatible in quantum theory. Although this question has no direct bearing on the experiment it must be answered one way or the other if a language for quantum theory is to be built up. Even if we should decide that the conjunction of S_1 and S_2 is not to be admitted as a sentence of the language of quantum theory, the question concerning the logical relationship between S_1 and S_2 still awaits an answer.

If we wish S_1 and S_2 to be compatible in quantum theory we must give up at least one of the principles from which their incompatibility has been derived. As the two main principles concerned are the existence of a potential energy and the conservation of total energy it is difficult to see how we can escape the conclusion drawn above without abandoning one of the principles on which quantum theory is based. In other words, S_1 and S_2 *are* incompatible within the present language of quantum theory, no matter how the syntax of this language may be specified with regard to other questions.

So far the solution of our paradox has not involved us in any logical difficulties; nor has it led to any alteration concerning the syntax of simultaneous measurement statements. However, we have not yet explored all implications of our simultaneous acceptance of (2.2') and of the incompatibility of S_1 and S_2; indeed, (2.1) is merely a consequence of, but not equivalent to, that incompatibility.

Let us now consider the logical relationship between S_1 and the following statement S_3:

> 's_0 is at t_0 in a state where the distance R has a definite value less than, or equal to, $r_{i\ \max}$'. $\hspace{2cm}$ (S$_3$)

In classical theory, S_1 and S_3 are compatible. Yet in quantum theory they prove incompatible if only the following postulate is accepted:

(2.P) If a [retrospective] measurement of a physical quantity X of a system gives the value x then, immediately before the measurement, the system was not in a state where the quantity X had a definite value different from x.

Indeed, let us accept S_1, and let a measurement of R be made which, in accordance with (2.2′), may give a value $r > r_{i\,max}$. We cannot assume that R had this value immediately before the measurement, since this (S_2) is incompatible with S_1. Hence only two possibilities remain: (i) R had a value less than, or equal to, $r_{i\,max}$ [i.e. S_3], or (ii) s_0 was not in a state where R had any definite value. If we assume (i), i.e. S_3, we violate the postulate (2.P). Hence, with (2.P) accepted, S_3 proves incompatible with S_1.

Now the postulate (2.P) may not be a necessary principle of theory construction. Still it is true, even in quantum physics, in all cases where it admits of a direct experimental test, i.e., when repreated measurements of one and the same quantity are made. The general validity of (2.P) in quantum theory is in fact implied by the mathemathical theory.

The possible incompatibility of a statement about one physical quantity with *any* statement about another quantity is a new and characteristic feature of quantum theory when compared with classical mechanics. Yet it is not entirely unknown to classical theory. As is well known the concept of temperature is defined only for a system in thermodynamic equilibrium. Hence any statement about the temperature of a thermodynamical system is incompatible with any statement characterizing that system as being in a state of non-equilibrium.

Sometimes the view is expressed, or tacitly held, that the statement 'a physical quantity has no value' is contradictory in itself. This is a mistake resulting from the incorrectness of our everyday mode of speech. In the latter the statement S_1, e.g., is usually replaced by

$$\text{'The energy of } s_0 \text{ at } t_0 \text{ is (or has the value) } w_i\text{'} \qquad (S_1')$$

or even

$$\text{'The value of the energy of } s_0 \text{ at } t_0 \text{ is } w_i\text{'}. \qquad (S_1'')$$

Grammatically, these sentences talk about a property of the energy (or the value of the energy) while they are actually about the physical system s_0. In a properly formalized language this mistake would automatically be avoided.

In the discussion of our paradox it is usually stated that the concept of distance cannot meaningfully be applied to the system s_0 if s_0 is an a state where the energy has a definite value. In a more precise form this statement is: the conjunction of the two sentences

$$\text{`}s_0 \text{ is at } t_0 \text{ in a state where the energy has a definite value } w_i\text{'} \tag{S_1}$$

and

$$\text{`}s_0 \text{ is at } t_0 \text{ in a state where the distance } R \text{ has a definite value } r\text{'} \tag{S_4}$$

is meaningless. (Since either of these sentences has a meaning we are not allowed to say 'one of them has no meaning if the other has'). This view is based on the argument that S_1 and S_4 cannot both be tested by direct measurement. Though true the argument is not conclusive. Most of the physical statements derive their meaning from experiment in a rather indirect and complicated way. Hence the question at issue can be decided only if a more precise definition of 'meaningless' is given, and even then the answer may depend on the way in which quantum theory is built up.

In this connection it must be emphasized that, contrary to what might be called vulgar logic, the possibility of two meaningful sentences forming a meaningless conjunction is not excluded by logic. In the case of statements like S_1 and S_4, the sentences which reduce the meaning of these statements to that of more elementary ones (e.g., statements used in experimental physics) in fact leave the meaning of the conjunction of S_1 and S_4 undetermined. This follows from the particular logical form of these reduction sentences in connection with the assumption that measurements of two different physical quantities cannot be performed simultaneously – an assumption that certainly holds in micro-physics.

What conclusions, if any, can be drawn from the previous discussion regarding the complementary mode of description suggested by N. BOHR[2]. First of all, what is the complementary mode of description? The answer, in Bohr's own words, is this:

That new mode of description is designated as complementary in the sense that any given application of classical concepts precludes the simultaneous use of other classical concepts which in a different connection are equally necessary for the elucidation of the phenomena.[3]

In slightly more technical terms we may re-state this explanation as follows:

the complementary mode of description is the usage of a language in which, according to its syntactic rules, *no simultaneous use* can be made of certain sentences belonging to that language.

The question remains: use for what? Two answers are possible: (i) use for *description*, and (ii) use for *prediction*. Of course these two interpretations do not exclude each other.

Now a description is made by means of sentences, or more precisely: by *one* sentence; [from the logical point of view the whole of this paper is one compound sentence, the full stops being short for 'and']. Hence, if we accept interpretation (i), the complementary mode of description is the usage of a language in which the formation of compound sentences out of given ones is restricted, i.e., the connection of two sentences of that language by means of 'and', 'or', etc., is not always again a sentence of that language.

We call such a language 'a *language with restricted connectibility*' or 'a language with inconnectible sentences'. For the sake of brevity we shall refer to such a language as a *complementary language in the formative sense* – 'formative', because only the rules of sentence formation are altered. Two inconnectible sentences may be called *F-complementary*.

Interpretation (ii) obviously concerns the rules of sentence transformation. We shall say a language is a *complementarity language in the transformative sense* if, and only if, the transformative rules of that language do not permit one to draw any consequences from certain pairs of sentences, except of course those that can be drawn from either sentence separately. The two sentences of such a pair may be called complementary (to each other) in the transformative sense, of briefly: *T-complementary*.

In principle, there is no connection between T-complementarity and F-complementarity, i.e., it is possible to build up a language in which two T-complementary sentences are not F-complementary and two F-complementary sentences are not T-complementary. However, if the conjunction of two sentences is not needed for a *description* of possible facts, those two sentences are certainly not needed simultaneously for a *prediction* of facts. Hence only those languages will be of practical interest to us in which any two F-complementary sentences are also T-complementary.

Now we come to our first conclusions regarding the connection between the paradoxes of quantum physics and the complementary mode of description.

(i) Since the *paradoxes of the first class* are concerned with only *two measurement statements* they *have no bearing on the question of T-complementarity.* The latter involves of course the consideration of at least three statements, two being possible premises and the third a possible conclusion.

(ii) The *paradoxes of the first class disappear if the condition of a measurement being made is duly accounted for in the formulation of the probability laws of quantum physics.*

This solution leads to the following consequence: if the principle (2.P) is to possess general validity in quantum theory any statement about one physical quantity must be considered *incompatible* with any statement about certain other physical quantities if both statements refer to the same system and the same instant. This constitutes a deviation from the particular syntax of classical mechanics but not from the general syntactic rules laid down by classical logic. In particular, it is *not necessary* to make any two of the statements concerned *F*-complementary, i.e. inconnectible, although of course it is permissible and not unreasonable to do so since the conjunction of two incompatible sentences gives a contradictory sentence, and contradictory sentences are not needed.

3. *An Instance of the Second Class*

Let us consider the following arrangement [cf. Figure 1]. On a solid frame two screens, Sc_1 and Sc_2, are erected parallel to each other. The first screen has two parallel narrow slits, S_1 and S_2, only a short distance apart from each other. The slits can be closed independently of one another by means of mechanisms fixed at the 'back' of the screen Sc_1, i.e., at the side facing the second screen Sc_2. In 'front' of Sc_1 is a 'shooting mechanism' from which a beam of elementary particles emanates, falling normally on the screen Sc_1. In order to avoid complications we assume that all particles have the same velocity and that the density in the beam is so small that interaction between the particles can be neglected. Sc_2 is a scintillation screen so that any particle hitting the screen, and the position of the hit on the screen, can be observed. For control purposes the flaps used for closing the slits are also painted on their 'front' sides with the same scintillating substance. For the sake of simplicity we assume that the intensity of the beam, i.e. the number of particles emitted per unit time, is constant throughout the period of the following experiments. This condition can easily be controlled by means of the scintillating flaps.

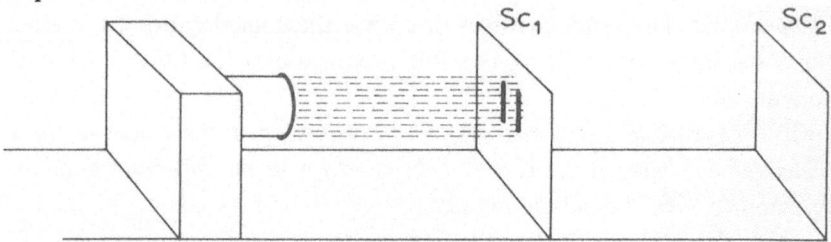

Fig. 1.

After making sure that no particle hits Sc_2 when the two slits are closed we make the following experiments:

(i) With S_1 open and S_2 closed, we observe the number and positions of hits on the screen Sc_2 during a certain time period, say one hour. The result can be represented by a distribution function

$$F_I(P)$$

where $P=(x, y)$ is the variable position on the screen Sc_2.

(ii) The experiment is repeated with S_1 closed and S_2 open. The distribution function in this case be

$$F_{II}(P).$$

(iii) The experiment is repeated, this time with both slits open. The distribution function found in this case be

$$F_{III}(P).$$

We should expect to find

$$F_{III} = F_I + F_{II}. \tag{3.1}$$

Actually, however, the experiment, according to quantum theory, gives a different result. While the total number of hits in the third case is still the sum of the number of hits in the two first cases:

$$N_{III} = N_I + N_{II}, \quad N = \int dP F(P) \tag{3.2}$$

the distribution is entirely different from (3.1):

$$F_{III} \neq F_{I} + F_{II} \qquad (3.3)$$

4. *Discussion of the Previous Paradox*

Let us first see on what argument the expectation (3.1) is based.
    Let

$$\text{prob}(0; S_1^p \wedge P) = f_{I}(P) \qquad (4.1)$$
$$\text{prob}(0; S_2^p \wedge P) = f_{II}(P) \qquad (4.2)$$

be the probability that a particle passes through the slit S_1 [or through the slit S_2, respectively] *and* hits the screen Sc_2 at P. ['0' symbolises the conditions under which the particles are emitted; '$\wedge$' stands for 'and']. Let

$$\text{prob}(0; [S_1^p \vee S_2^p] \wedge P) = f_{III}(P) \qquad (4.3)$$

be the probability that a particle passes through S_1 or S_2 *and* hits the screen Sc_2 at P; ['$\vee$' stands for 'or']. Applying to (4.3) the general addition theorem of the theory of probability[4] we obtain

$$\begin{aligned}
\text{prob}(0; [S_1^p \vee S_2^p] \wedge P) = {} & \text{prob}(0; S_1^p \wedge P) + \\
& + \text{prob}(0; S_2^p \wedge P) - \\
& - \text{prob}(0; S_1^p \wedge S_2^p \wedge P),
\end{aligned}$$
$$(4.4)$$

after having used the logical equivalence

$$[X \vee Y] \wedge Z = [X \wedge Z] \vee [Y \wedge Z].$$

Thus we get

$$f_{III}(P) = f_{I}(P) + f_{II}(P) - \text{prob}(0; S_1^p \wedge S_2^p \wedge P). \qquad (4.4)$$

If we admit that a particle cannot pass through both slits simultaneously:

$$\text{not}(S_1^p \wedge S_2^p), \qquad (4.5)$$

the last term in (4.4) is zero and we have

$$f_{III}(P) = f_{I}(P) + f_{II}(P).$$

This is equivalent to (3.1) since the distribution functions F differ from the probability functions f only by a common factor.

Our expectation (3.1) is thus based on the assumption (4.5), and it seems that the only way in which to account for a deviation from (3.1) is to abandon (4.5).

The abandonment of (4.5) is suggested by what are called the wave-properties of the elementary particles. Indeed, if we assume that a particle, like a wave, can pass partly through one and partly through the other slit the last term in (4.4) will produce a deviation from the additive law (4.6), or (3.1).

At first sight the assumption

$$\text{prob}(0;\, S_1^p \wedge S_2^p \wedge P) \neq 0 \tag{+}$$

seems to be incompatible with the experimental law $N_{III} = N_I + N_{II}$, as the integration of (4.4) over P seems to give a number N_{III} less than the sum of N_I and N_{II}. This argument, however, is not correct. If S_1^p does not exclude S_2^p, then f_1 counts not only the particles passing entirely through S_1 but also those passing partly through S_1 and partly through S_2. [The subtractive term in (4.4) simply means that no particle is counted twice.] It follows that f_1 can no longer be identified with F_I since F_I is won with S_2 closed whereas f_1, according to its definition, does not involve this condition. Only if (4.5) is fulfilled does S_1^p describe the condition of the first experiment ('if (part of) a particle passes through S_1 the whole of it passes through S_1'), and hence only then can f_1 be identified with F_I.

The assumption that a particle can pass partly through S_1 and partly through S_2 is difficult to reconcile with another experimental fact not yet mentioned. This is the fact that the number of hits on the closed slit is independent of whether the other slit is closed or not. This follows from quantum theory and could easily be tested by means of the control flaps mentioned above. The difficulty is this. On the one hand, the number of particles *hitting* the closed slit, S_1, say, is independent of whether S_2 is open or closed. On the other hand, the number of whole particles *passing* through S_1 depends on whether S_2 is open or closed. Indeed, if both slits are open some, if not all, particles divide themselves for the passage through, and hence the number of whole particles passing through one slit decreases. This difficulty is underlined by the following consideration. Let N_1 be the number of particles passing through S_1 in case S_2 is closed. Let n_1 be the number of particles passing (wholly or partly) through S_1 in case S_2 is open. Now the opening of S_2 might give some of the N_1 particles a change to pass, partly or entirely, through S_2. *Hence we should expect n_1 to be less than, or at most equal to, N_1.* ('equal' because n_1 includes also those particles of which only a part passes through S_1). This, however, is incompatible with the assumption that the number of particles going partly through S_1 and partly through S_2, n_{12}^p, say, is not zero. Indeed, the definition of our numbers is

$$N_1 = C \operatorname{prob}(0 \wedge S_1 \wedge \bar{S}_2; S_1^p) \tag{4.71}$$
$$N_2 = C \operatorname{prob}(0 \wedge \bar{S}_1 \wedge S_2; S_2^p) \tag{4.72}$$
$$n_1 = C \operatorname{prob}(0 \wedge S_1 \wedge S_2; S_1^p) \tag{4.81}$$
$$n_2 = C \operatorname{prob}(0 \wedge S_1 \wedge S_2; S_2^p) \tag{4.82}$$
$$n_{12} = C \operatorname{prob}(0 \wedge S_1 \wedge S_2; S_1^p \wedge S_2^p) \tag{4.83}$$

where 'S_1' and '$\bar{S}_1$' represent the conditions 'S_1 is open' and 'S_1 is closed', respectively; C is a common proportionality factor which depends solely on the intensity of the beam and the time period chosen. Now the total number of particles passing through the screen Sc$_1$ when both slits are open is given by

$$n = C \operatorname{prob}(0 \wedge S_1 \wedge S_2; S_1^p \vee S_2^p). \tag{4.9}$$

Using the general addition theorem we get

$$n = n_1 + n_2 - n_{12}. \tag{4.10}$$

On the other hand, this number, as already mentioned, is equal to $N_1 + N_2$:

$$N_1 + N_2 = n_1 + n_2 - n_{12}. \tag{4.11}$$

Hence, if n_{12} is not zero, n_1 must be greater than N_1 for reasons of symmetry, contrary to what we were led to expect by the consideration above.

Finally, the question whether a particle can pass partly through S_1 and partly through S_2 can, in a certain sense, be decided by experiment. Of course, we must not use an arrangement which, in principle, would allow us to decide whether a particle passes through one particular slit; any such arrangement might exclude the possibility of a particle going through both slits and hence prejudice the outcome of the experiment. In fact, any such arrangement makes the phenomenon of 'interference' on the screen Sc_2 disappear and re-establishes the simple additivity. However, with counter arrangements as used in work on cosmic radiation it is possible to count simultaneous passages through both slits, and only such passages. It might be objected that these counters count the simultaneous passages of *two* particles, not of two parts of *one* particle. However, if we use counters which respond to energy passing through, and not to electric charge, this objection is no longer convincing. If the statement that a particle passes partly through the one slit and partly through the other is to have any meaning at all it can only mean that one part of its proper energy passes through the one and the other part through the other slit, as would be the case with a wave. As is well known, the experimental decision, according to quantum theory, is in the negative.

To sum up: *it is not possible to solve the paradox by the assumption that (4.5) does not hold*, i.e., that a particle can partly pass through one and partly through the other slit. It should be emphasized that this conclusion is based on experimental facts as predicted by quantum theory and the calculus of probability only. No use is made of the notion of causality, or of any properties of 'particles' which are contrary to quantum theory. Whether we say 'particle', or 'wave', or 'something', does not matter for our argument. *There is only one way left to solve the paradox, namely to abandon the tacit assumption that all the probabilities used in our argument exist*. That a probability, 'prob$(X; Y)$', say, exists means that

$$\text{prob}(X; Y) = z \quad (0 \leqslant z \leqslant 1) \tag{$++$}$$

is a sentence. A necessary condition for $(++)$ to be a sentence is that 'X' and 'Y' are predicates. Hence if we abandon the assumption that all the argument expressions occurring above are predicates the paradox can no longer be produced. Since there can be no doubt that each of the symbols '0', 'S^p', 'P' is a predicate *the assumption to be abandoned is that the connection of these predicates gives a predicate again*.

Thus we arrive at the complementary mode of description, the latter understood both in the formative and the transformative sense. The following remarks will elucidate this statement.

The experience upon which the suggestion of the complementary mode of description is based is two-fold. First, we have the impossibility of simultaneous measurements of certain physical quantities. This fact has no bearing on the question of T-complementarity and does not compel us to accept F-complementarity. Second, if two quantities such as momentum and position are measured, the one after the other, the probability for a certain result of a third measurement does not depend on the outcome of the first measurement; in other words: the first two measurements need not simultaneously be used for a prediction. We say 'need not' instead of 'cannot' because the latter is justified only if the simultaneous use would lead to contradictions. That it does lead to contradictions has been shown above, although in a somewhat indirect way. In our formulae the contestable connection of predicates does not occur in the first argument place of 'prob' but in the second, i.e. it occurs, not as a 'premise', but as a probable 'conclusion'. However, within the calculus of probability this distinction is without significance since both cases are connected by the general product theorem:

$$\text{prob}\,(0;\,S^p \wedge P) = \text{prob}\,(0;\,S^p) \times \text{prob}\,(0 \wedge S^p;\,P)\,.$$

It was easier to discuss the paradox in terms having the form of the left-hand probability, but if this probability does not exist the second probability on the right-hand side does not exist either.

In this connection a particular point deserves emphasis. It has been pointed out by PAULI[5] that the phenomenon of 'interference', i.e., the deviation from the additive law, disappears if the slits are controlled in a way that enables us to say which of the two slits the particle has passed through. Yet even a control that would merely enable us to say whether or not a particle has passed through one and/or the other slit would make the deviation disappear. Indeed, as far as the question of complementarity is concerned, the disjunction of two statements about the *same* physical quantity behaves in the same way as either component.

II. COMPLEMENTARITY AND THE MATHEMATICAL THEORY

5. *The Probability Function of Quantum Theory*

It is usually said that the mathematical criterion for two physical quantities to be complementary is the non-commutability of the two operators representing the two quantities. What however is meant by saying that two physical quantities are complementary? If the answer is 'the measurement of the one quantity makes it impossible to measure the other quantity', the above statement is not always true. Indeed, let A and B be the two operators representing two physical quantities A and B at a certain time t. B at the time $t+\tau$ is then represented by the operator

$$B^\tau = U^\tau B (U^\tau)^{-1} \qquad (5.1)$$

where U^τ is a unitary operator given by

$$U^\tau = \exp (iH\tau/\hbar) \qquad (5.2)$$

H being the Hamiltonian operator and $\hbar$ being Planck's constant divided by 2π. It is easy to prove that if A and B do not commute A and B^τ do not commute either. Moreover, if A does not commute with H even A and A^τ do not commute. Yet the quantities A and B^τ can be measured both since the two measurements can be performed at different times, t and $t+\tau$, respectively.

It follows that if the non-commutability is to be a criterion of complementarity the latter cannot be restricted to simultaneous measurements and hence must have a significance independent of the impossibility of simultaneous measurents. This is in full agreement with our result that the so-called paradoxes concerning simultaneous measurement statements do not compel us to adopt the complementary mode of description.

The simplest, and most efficient, way to use the mathematical formalism of quantum theory for the question of complementarity is to analyze the mathematical expression by which the probabilities of quantum theory are determined. Indeed, the syntax of the term 'probability' is given to us by the modern theory of probability, both as to the formative and as to the transformative rules. On the other hand, quantum theory contains, so to speak, its own theory of probability. (Unlike the statistical theories of classical physics, quantum theory does not start from certain probability assumptions from which other probability statements are derived by explicit use of the calculus of probability; instead, we have a coherent mathematical formalism to which a probabilistic interpretation is attached; hence *quantum theory contains a theory of probability implicitly*). By comparison we can, therefore, find both the syntactical significance of the relationship of complementarity and the answer to the question to what expressions that relationship applies.

One thing can be said beforehand. The transformative rules for 'probability' (the so-called postulates or axioms of the theory of probability) can be proved on the basis of any frequency interpretation, be the latter finite or infinite. Hence we can expect deviations from the classical formative rules only. This does not mean that no transformative significance attaches to a deviation from the classical formative rules: the rules of sentence transformation presuppose the rules of sentence formation, and hence an alteration of the latter affects the *range* to which the transformative rules can be applied, though not the transformative rules themselves.

For the following discussion it proves convenient to write

$$`Z_{\Delta x}^{X;\tau}(s, t)` \quad \text{for} \quad \text{'The system } s \text{ is at the time } t+\tau \text{ in a state}$$

where the value of the quantity x lies in the interval Δx'   (5.3)

The probability $\text{prob}(Z_{\Delta a}^{A}; Z_{\Delta b}^{B;\tau})$, according to quantum theory, is given by the following mathematical expression:

$$\text{prob}(Z_{\Delta a}^{A}; Z_{\Delta b}^{B;\tau}) = F_P(E_{\Delta a}^{A}; E_{\Delta b}^{B;\tau}) \underset{\text{df}}{=} Tr(E_{\Delta a}^{A} E_{\Delta b}^{B;\tau})/Tr(E_{\Delta a}^{A}).$$

(5.4)

The E's are projectors (projection operators) in Hilbert space, defined as follows. If the operator X is written as

$$X = \int dx E^X(x) \quad \text{i.e.} \quad (Xf, g) = \int x \, d(E^X(x) f, g)$$

(5.P)

where f and g are any two elements of the Hilbert space, (f, g) being the 'inner product' of f and g, then

$$E_{\Delta x}^{X} = E^X(x'') - E^X(x') \quad \text{if} \quad \Delta x = (x', x''].$$

(5.5)

Tr, called the 'trace', is a numerical function defined by

$$Tr \, Y = \sum(Yf_i, f_i)$$

(5.6)

where the f_i are any complete set of normalized functions

$$(f_i, f_k) = \begin{cases} 0 & \text{for} \quad i \neq k \\ 1 & \text{for} \quad i = k \end{cases}.$$

(5.7)

Finally, we have used the definition

$$E_{\Delta x}^{X,\,\tau} = U^\tau E_{\Delta x}^X (U^\tau)^{-1} \tag{5.8}$$

where U^τ is defined by (5.2).

From (5.4) it can be seen that, in quantum theory, we have a one-one correspondence between predicates of the form $Z_{\Delta x}^X$ and projection operators. Hence the connective terms 'and', 'or', used for building compound predicates, must correspond to the connective symbols used for building compound projection operators. If the projection operators E and E' correspond to the predicates Z and Z', respectively, the correspondence between the connective terms is given by the following table:

Compound Predicates	Compound Projection Operators
$Z \overset{.}{\wedge} Z'$	$E\,E'$
$Z \overset{.}{\vee} Z'$	$E \overset{.}{+} E' = E + E' - E\,E'$

$$\tag{5.9}$$

That this correspondence is the right one can be proved in two ways. First, it gives the right probabilities. Second, it makes the function F_P defined in (5.4) fulfil the functional equations of the theory of probability, i.e., the general product theorem and the general addition theorem. For the proof of the latter use has to be made of the fact that Tr is an additive function.

Now we come to the conclusion. $E\,E'$ and $E \overset{.}{+} E'$ are again projection operators if, and only if, $E\,E' = E'\,E$. Incidentally, this is also the necessary and sufficient condition for the correspondence to be a one-one correspondence. Now a principle as simple as the one-one correspondence between projection operators and predicates can hardly be expected to break down if it is extended to compound predicates. Thus we may draw the conclusion that the compound predicates do not exist if the corresponding operators are not projection operators. In other words: *the criterion for two predicates to be complementary (inconnectible) is given by the non-commutability of the two corresponding projectors.*

It is noteworthy that this conclusion can be confirmed by another consideration not involving the above postulate of a *general* one-one correspondence between predicates and projectors. We may, on trial, assume that '$Z_1 \wedge Z_2$' is a predicate even if $E_1 E_2$ is not a projector, i.e., if $E_1 E_2 \neq E_2 E_1$. Now let us consider the expression

$$\text{prob}(Z; Z_1 \overset{.}{\wedge} Z_2) = F_P(E; E_1 E_2) = Tr(E E_1 E_2)/Tr(E)$$

(§)

and let us apply the following *theorem* which can be proved easily:

THEOREM: If E, E_1', E_2'' are projectors in a Hilbert space it is always

$$Tr(E E_1 E_2) = Tr(E_2 E E_1) = Tr(E_1 E_2 E) = c$$
$$Tr(E E_2 E_1) = Tr(E_1 E E_2) = Tr(E_2 E_1 E) = \bar{c}$$

where $\bar{c}$ is the conjugate complex of c. $c = \bar{c}$ if, and only if, at least two of the three projectors commute with one another. Hence, if none of the operators E, E_1, E_2 commute we obtain a complex number for the probability in (§), i.e., that probability does not exist.

At first sight it might look surprising that a further condition of non-commutability has to be added to the original one in order to make the numerical expression $F_P(E; E_1 E_2)$ assume a complex value. However, it must not be forgotten that, according to the general product theorem of the theory of probability, this numerical expression must give the value of two other probability expressions, namely

$$\text{prob}(Z; Z_1)\, \text{prob}(Z \overset{.}{\wedge} Z_1; Z_2)$$

(§§)

and

$$\text{prob}(Z; Z_2)\, \text{prob}(Z \overset{.}{\wedge} Z_2; Z_1)$$

(§§§)

[As already remarked, we have $F_P(E; E_1 E_2) = F_P(E; E_1) F_P(E E_1; E_2) = = F_P(E; E_2) F_P(E E_2; E_1)$.] Hence, if one of the compound predicates '$Z \overset{.}{\wedge} Z_1$', '$Z \overset{.}{\wedge} Z_2$', '$Z_1 \wedge Z_2$', and therefore one of the three probabilities (§§), (§§§), (§), exists, that numerical expression must be a real number.

Thus the mathematical formalism of quantum theory is in complete agreement with the results obtained in the discussion of the paradoxes. Moreover, a better understanding of the mathematical formalism is attained. It has always been difficult to explain why the operators representing the physical quantities must be 'hyper-maximal', i.e., must be capable of the representation (5.P) involving projectors.

From our point of view this is very easy to understand: the calculus of the projectors is isomorphic to the modified calculus of predicates which emerges from the classical one if the condition of unrestricted connectibility is abandoned. Furthermore, a clear answer can now be given to the question whether the use of complex-valued functions is essential to quantum theory or merely a technical device not actually needed. If we were to use real-valued function, i.e. if we were to replace the Hilbert space with its unitary metric by a vector space with Euclidean metric, the function F_P in (§) would yield real numbers even if none of the three corresponding probabilities existed.

NOTES

* Published for the first time. The paper is undated but must have been written between 1939 and 1945.
[1] M. Strauss, 'Zur Begründung der Statistischen Transformationstheorie der Quantenphysik', *Sitz. Ber. Berl. Akad. Wiss., Phys.-Math. Kl.* **27** (1936), 90–113. [Chapter XVI of this volume.]
[2] G. Birkhoff and J. von Neumann, 'The Logic of Quantum Mechanics', *Ann. of Math.* **37** (1936), 823.
[3] N. Bohr, *Atomic Theory and the Description of Nature.* Cambridge 1934.
[4] Cf., e.g., H. Reichenbach, *Wahrscheinlichkeitslehre*, Leiden 1935.
[5] W. Pauli, 'Allgemeine Prinzipien der Quantenmechanik' in *Hd. Physik* 2nd edition (ed. by H. Geiger and K. Scheel), vol. **24**, Part 1, Berlin 1933.

CHAPTER XVIII

QUANTUM THEORY AND LOGIC*

The continued discussion about the so-called logical problems of quantum theory (QT) [cf., *e.g.*, Reichenbach's book, recent issues of *Dialectica* and *Philosophy of Science*] is due to the absence of a recognized physical axiomatics to which these problems could be referred for settlement. The failure of various attempts to reconstruct QT on the basis of simple physical postulates is a sign that a complete understanding of QT has not yet been achieved, or else that something entirely new is involved in such a reconstruction.

The reconstruction of the basic features of mathematical QT, carried out in 1936 by the author, is based on what may be described as a translation of Bohr's conception of complementarity into the language of logical syntax where it gives rise to a new 'logic' characterized by restricted sentential connectibility. The main contentions of the author's lecture were (i) that this complementarity logic is absolutely essential in any rational reconstruction of QT, and (ii) that it accounts for the most general features of our experience in the field of quantum physics in exactly the same sense in which non-Euclidean geometry accounts for the basic feature of gravitation, *i.e.* the [local] equivalence of gravitation and acceleration, or in which the Lorentz-transformation accounts for the basic fact in kinematics, viz., the existence of a finite limiting velocity.

The 'most general feature of quantum physical experience' may be extracted either from an analysis of QT itself or from that of typical quantum phenomena. It consists (i) in the statistical character of all well-defined predictions, and (ii) in the relationship of *general complementarity* which holds between the results of any two actual measurements or observations (other than those of a constant quantity), and which may be stated thus: *the results of two measurements cannot be used simultaneously in the same prognostic or retrospective argument if physically correct conclusions are to be drawn.* In the non-relativistic theory this reduces to the simpler statement that of two consecutive measurements only the first can be used for a retrospective analysis while only the second can be used for prognosis. The much discussed impossibility of simultaneous measurement and indeed of the definition of simultaneous attributes (special complementarity), far from being the root of indeterminacy in quantum physics, turns out to be a logical consequence of the relationship of general complementarity.

In discussing methods of theory construction the author had already suggested that the secret of progress in theoretical physics lies in the choice of a technical language which by its very syntax accounts for the most general features of our physical experience, a secret first revealed in the construction of Einstein's theory of gravitation. Using this method, and accordingly formulating the relationship of general complementarity as a syntactic rule for the quantum theoretical language, one obtains directly the above-mentioned 'complementarity logic' (restricted sentential connectibility) the calculus of which is isomorphic to the calculus of projection operators used in the general formulation of QT. Moreover, a consideration of compound probability expressions such as occur in the general theorem for the multiplication of probabilities then shows that the projection operators have to be Hermitean (and not real) since otherwise the mathematical theory would give an answer to questions to which, according to the principle of general complementarity, there is no definite answer in terms of probabilities. *The use of complex-valued probability amplitudes in the Schroedinger representation of the theory is thus shown to be a direct consequence of the principle of complementarity.*

While it would be tempting to discuss the ideas advanced by other workers in this field the results of such a discussion would have an essentially negative character since none of these ideas is powerful enough to result in a reconstruction of QT on an empirical basis.

The reconstruction of QT outlined above leads to a separation of those features of present QT which are fundamental (projection operators in a space of unitary metric) and those which could be changed without changing the foundations (existence of a Hamiltonian, localizable fields) and which at present are determined by the principle of correspondence. (Among the quantum theories generally accepted at present the QT of the spinor fields is the only one which goes beyond this principle. However, it does not violate this principle since the latter is inapplicable to spinor fields (the notion of a spinor has no place in classical theory).)

Unless the principle of correspondence is ill-applied in present QT, a break with the restrictive requirements of this principle appears to be necessary for any substantial progress in the theory of elementary particles. On the other hand, the principle of complementarity, and with it the fundamental features of mathematical QT, do not appear capable of any generalization and must therefore be expected to stay.

From the general point of view of theoretical logic (theory of theory construction), the most interesting feature of QT is perhaps not that it involves a sentential calculus different from any one invented by the professional logicians, but rather that this calculus together with its interpreting language (*i.e.*, "the language of experimental physics supplemented with the terminology of classical theory" (Bohr)) constitutes a language-system of a syntacto-semantic type more general than any of the types considered hitherto. Indeed, only in the case of special complementarity would the conjunction of two inconnectible sentences represent a meaningless statement, while in the case of general complementarity the rule of inconnectibility cannot be 'justified' on such simple grounds of semantics. By implication, then, Carnap's principle of tolerance has been extended from the realm of syntax to the wider field of syntacto-semantics.

In his concluding remarks the author expressed the opinion that the task of the logic of science lies in the future rather than in the past, and that the theory of theory construction, in its relation to theoretical physics, will develop in a way similar to the way in which theoretical physics itself has developed in relation to experimental physics – from a tool of analysis and synthesis *post factum* through a period of close mutual collaboration to an operative role in the progress of science.

NOTES

* Abstract of paper read on February 20, 1950, at a meeting of the *Philosophy of Science Group* of the *British Society for the History of Science*. Reprinted from the *Bulletin of the British Society for the History of Science*, Vol. 1, No. 4 (October 1950), 99–101.
[1] 'Zur Begründung der Statistischen Transformationstheorie der Quantenphysik', *Sitzungsberichte der Preussischen Akademie der Wissenschaften zu Berlin, phys.-math. Kl.*, **27** (1936), 90–113.

CHAPTER XIX

FOUNDATIONS OF QUANTUM MECHANICS*

1. Introduction and summary

When [in 1905] the c-theory [Special Theory of Relativity] was born, both the mathematical formalism and its physical interpretation were established simultaneously; merely some questions of physical logic and axiomatics remained to be clarified. The story is quite different in the case of h-theory. To start with, two apparently quite different mathematical theories emerged, known as 'wave mechanics' (Schroedinger) and 'matrix mechanics' (Heisenberg-Born-Jordan), respectively. The underlying physical conceptions and, hence, the first physical interpretations were entirely different: Schroedinger believed he had reduced the quantum phenomena to a classical eigenvalue problem of the sort known from the theory of oscillations while Heisenberg-Born-Jordan understood their theory as a fundamental generalization of classical mechanics satisfying Bohr's principle of correspondence. The progress achieved in the following time consisted of three main steps.

First, after Schroedinger had shown how one of the two mathematical theories could be translated into the other, von Neumann showed that the two mathematical formalisms were *isomorphic*, to wit, different realizations [models₂] of the axiomatically defined [abstract] *Hilbert space*. To the Schroedinger function $\psi(q)$ corresponds a line or column U_a of an infinite quadratic unitary matrix [and vice versa] and to the [normalisation] condition

$$\int \int \int \psi^*\psi \, d^3q = 1$$

corresponds the equation

$$\sum_a U_a^* U_a = 1.$$

Both $\psi(q)$ and U_a are *normalized vectors in Hilbert space*. At the same time it became clear that the difference between the two versions of the theory resides merely in the fact that different systems of coordinates in Hilbert space are used as preferential systems. The study of these questions led to the so-called *quantum mechanical transformation theory* (Jordan-Wigner, Dirac), which is an exact counterpart to the kinematical transformation theory of *c*-mechanics. While the kinematical transformations [so-called Lorentz transformations] of *c*-theory mean the transition from one space-time frame of reference to another one [within a so-called uniform motion equivalence], the unitary transformations of quantum theory represent the 'transition' between different measuring arrangements [or rather: different external conditions suitable] for measuring noncommensurable quantities represented by [noncommuting] Hermitean operators in Hilbert space. While it was clear from the beginning that the *eigenvalues* of such an operator represent the possible values of the quantity concerned, the physical meaning of these unitary transformations (rotations in Hilbert space) remained a question to be clarified.

A first answer [in the right direction] was the *statistical interpretation* given by M. Born. Its general formulation is as follows. Let

$$U_{ab}^{AB} = \int (\psi_a^A)^* \, \psi_b^B \, dq \; = \; \langle a \mid b \rangle \tag{1}$$

$$\text{BORN} \qquad \text{SCHROEDINGER} \qquad \text{DIRAC}$$

be the [elements of] the matrix of unitary transformation from the system of eigenvectors of A with eigenvalues a to the system of eigenvectors of B with eigenvalues b; then:

$$|U_{ab}^{AB}|^2 = |\langle a \mid b \rangle|^2 \tag{2}$$

is the [value of the] probability that the quantity [represented by] B has the value b, if [and when] [the quantity represented by] A has the value a. According to this interpretation quantum mechanics would be a *statistical theory in the classical meaning of this word*; its statements would refer in principle to statistical ensembles of like systems only [but not to single systems and elementary processes].

The third [and final] step consisted in replacing the statistical by the *probabilistic [stochastic] interpretation*. According to this interpretation, the words

'*has* the value b'

in the above formulation have to be replaced by

> '*takes* the value *b* when an interaction [of the given system with external conditions] comes into play that corresponds to a measurement of [the quantity represented by] *B*'.

Thereby, the quantum mechanical probabilities refer to *transitions* and *not* to statistical *distributions*.

It is now clear how the logical reconstruction of the theory has to proceed. Just as in *c*-theory the concept of [constant] *velocity* has to be analysed and axiomatised, so here the concept of [transition] *probability* has to be analysed and axiomatised. Just as the Lorentz transformation appeared as [true irreducible] representation [in *x-y-z-t*-space] of the [abstract] velocity group, the transformation theory of quantum mechanics will appear as a representation [or model$_2$] of the axiomatic [abstract] theory of [transition] probabilities. On this basis the quantum mechanical 'law of motion' is obtained in a way analogous to the way in which *c*-dynamics is obtained [on the basis of *c*-kinematics]: instead of Lorentz invariance we have to demand here invariance under the group of unitary transformations, and instead of the limit relation for $c \to \infty$ we have to demand here an analogous relation for $h \to 0$.

As all other analogies, that between the two transformation theories is also incomplete: while the [invariant] quantity *c* appears already in the transformation equations, the [invariant] quantity *h* only appears in the next step which introduces dynamical quantities. This is of fundamental importance for the logical structure of the theory and its proper understanding. It implies that the transition $h \to 0$ cannot be carried out for the underlying mathematical formalism (Hilbert space). [Physically,] this has to do with the probabilistic [stochastic] character of *h*-theory: In the limit $h \to 0$ all operators will commute (become *c*-numbers, in Dirac' terminology) and consequently all transition probabilities will cease to exist. Thus classical mechanics appears far more *conceptually degenerate* from the standpoint of *h*-mechanics than it does from the standpoint of *c*-mechanics. This state of affairs is most clearly expressed in terms of formal logic: *h*-mechanics is based not only on a different mathematical formalism but also on a different predicate logic (complementarity logic); only the latter, but not the mathematical formalism, is a proper generalisation of classical theory.

2. STATISTICAL PROBABILITY

The classical (statistical) concept of probability may be defined as that concept which occurs in statements of the following form:

> The probability that a subject with property E_1 also has the property E_2 is equal to p.

For this we write

$$\text{prob}_1(E_1; E_2) = p. \tag{1}$$

If we take casting dice as an example, 'E_1' would be the predicate 'properly cast' and 'E_2' would stand for 'lying on the table with a six, say, on the top face'. Now predicates may be negated and [if of the same syntacto-semantic type] combined by 'and' ($\wedge$) and 'or' ($\vee$) [to give new predicates of the same type]. The rules of the classical calculus of predicates or classes apply. Thus, the arguments of the classical probability functor 'prob$_1$' are elements of a *Boolean algebra* (also called *distributive orthocomplemented lattice*). The axioms of classical prob theory are conditions for the function prob$_1$ [the rules of Boolean algebra being taken for granted]. These axioms are chosen such that they permit the *interpretation of prob$_1$ as relative frequency*; in this interpretation the axioms are tautologically satisfied or become mathematical identities.

The frequency [statistical] interpretation has the form

$$\text{prob}_1(E_1; E_2) = \frac{f(E_1 \wedge E_2)}{f(E_1)}; \tag{2}$$

it reduces the two-place function prob$_1$ to the one-place function f. $f(E)$ is the number of objects with the property E. From this meaning of 'f' and the meaning of '$\vee$' it follows that

$$f(E_1 \vee E_2) = f(E_1) + f(E_2) - f(E_1 \wedge E_2). \tag{3}$$

Any function satisfying this functional equation is called an *additive function over a Boolean algebra*.

It is easy to prove that (2) together with (3) is equivalent to the axioms of the classical theory of probability as formulated, e.g., by H. Reichenbach in 1932. In this proof the meaning of f does not play any role. Thus, from the mathematical point of view the statistical theory of probability is identical with the *theory of additive functions over a Boolean algebra*.

This fundamental result is due to Kolmogoroff (1933). In his work a *set system* (consisting of all subsets of a given set) is used as model$_2$ of the Boolean algebra. In contrast to this we maintain the view that the arguments of prob$_1$ and f are predicates: only this view can be taken over, in a generalized form, into quantum mechanics.

3. REACTIVE PROPERTIES–COMPLEMENTARITY LOGIC

As shown above, the classical (statistical) theory of probability rests on the classical calculus of predicates. Hence it presupposes that the logical conjunction '$E_1 \wedge E_2$' is meaningful if 'E_1' and 'E_2' are meaningful. This presupposition need not however be satisfied for predicates representing reactive ['dispositional'] properties. Such predicates will be represented in the following by 'X', 'Y', 'Z'.

A typical predicate of this kind is 'soluable in water'. As this example shows, such predicates cannot be defined explicitly; they admit of only a partial definition of the form

$$E_1 \wedge X \equiv E_2 \tag{1}$$

(in the example: $E_1 = $ is in water, $E_2 = $ is dissolving in water).

If a second predicate of this kind (say: 'soluable in alcohol') with the definition

$$E_3 \wedge Y \equiv E_4 \tag{2}$$

is considered, the conjunction '$X \wedge Y$' is merely defined by

$$(E_1 \wedge E_3) \wedge (E \wedge Y) \equiv E_2 \wedge E_4. \tag{3}$$

Now it may happen that the properties E_1 and E_3, and the properties E_2 and E_4, exclude one another so that

$$E_1 \wedge E_3 \equiv E_2 \wedge E_4 \equiv O \tag{4}$$

where 'O' stands for the contradiction. In this case (3) is identically satisfied, i.e., $X \wedge Y$ remains *completely undefined* [and hence possibly without meaning][1]. Thus, classical logic contains the inherent possibility of annulling itself.

It is precisely this possibility that has become a reality in quantum mechanics.

Following N. Bohr, predicates the logical conjunction of which is un-defined and hence without meaning will be called *complementary* (to each other) (in the strict sense). The corresponding properties will also be called complementary (to each other) or *incommensurable*.

In view of the logical identity

$$X \vee Y \equiv \overline{\overline{X} \wedge \overline{Y}}$$

it follows that for complementary predicates the disjunction is likewise undefined.

From the definition given above it follows that complementarity is an irreflexive symmetrical relation between predicates while nothing follows concerning the question of its being transitive or intransitive.

Since expressions without meaning should not occur in scientific language, the formation of compound predicates out of complementary predicates is to be forbidden by a syntactical rule. It then follows: *complementary predicates are inconnectible*. Inconnectibility thus appears as the *syntactic formulation of complementarity*. The sentential and predicate calculus resulting from the admission of inconnectible pre-dicates [or rather from *restricted* connectibility] is called *complementarity logic*.

In complementarity logic the universal connectibility of predicates and sentences is abolished. Hence its algebraic structure is no longer that of a Boolean algebra. However, since the relation of complementarity is not [by definition] transitive, complementarity logic in general admits connectible predicates. Hence its algebraic structure is called *partial Boolean*.

4. THEORY OF PROBABILITY ON THE BASIS OF COMPLEMENTARITY LOGIC

Logical connectibility is replaced in quantum mechanics by transition probability. If X and Y are complementary properties there always exists a number p such that

$$\mathrm{prob}_2(X; Y) = p. \tag{1}$$

The question now arising is this: what calculus applies to the transition probability prob_2?

The answer is not difficult to find: we must have a prob calculus on the basis of complementarity logic, i.e., the arguments of $prob_2$ belong to a partial Boolean algebra. As far as these arguments are connectible the rules of the ordinary calculus of probability must apply. This leads to the result: *the rules of the classical theory of probability still hold provided the arguments (predicates) exist.*

The problem is thus reduced to that of finding a *mathematical model₂ of complementarity logic.* The solution found by quantum mechanics may be stated in generalized form as follows: *every predicate of complementarity logic is represented mathematically by a projection operator in a linear vector space.* The connectives for predicates and operators, respectively, are correlated as follows:

$$X \leftrightarrow P_X$$
$$X \overset{.}{\wedge} Y \leftrightarrow P_X P_Y \qquad \text{(a)}$$
$$X \overset{.}{\vee} Y \leftrightarrow P_X + P_Y - P_X P_Y \quad \text{(b)} \qquad (2)$$
$$\overset{.}{X} \leftrightarrow I - P_X \qquad \text{(c)}$$

With this scheme a *mathematical criterion* of complementarity is found: *two predicates are complementary iff the projection operators representing them do not commute.* Indeed, if two projection operators do not commute their product is no longer a projection operator and hence a corresponding [compound] predicate does not exist.

The remaining problem is that of finding the *general solution of the functional equations* for $prob_2$ in terms of projection operators. Obviously, the solution must again have the form

$$prob_2(X; Y) = \frac{S(P_X P_Y)}{S(P_X)} \qquad (3)$$

where S is a real-valued additive function:

$$S(P_X + P_Y) = S(P_X) + S(P_Y). \qquad (4)$$

The only function satisfying this condition is the *trace* (Tr), which may be defined as the sum of the eigenvalues of the argument operator.

Hence the complete solution of our problem is given in general form by the equation

$$prob_2(X; Y) = \frac{Tr(P_X P_Y)}{Tr\, P_X} \qquad (5)$$

This is the formula replacing Kolmogoroff's formula (2). At the same time it is the *fundamental formula of quantum mechanics*.

The space in which the projection operators are defined is the *representation space of complementarity logic*. As long as nothing else is known, number of dimensions and metric of the representation space are arbitrary, i.e., all vector spaces have equal rights to be considered. However, if (5) is taken into account it is easy to show that *unitary* (and not *real*) metric has to be chosen, i.e., the vectors of the representation space have to be *complex* quantities. Indeed, if the representation space were a *real* vector space, (5) would yield a real number for non-existing probabilities! The use of *complex*-valued functions (vectors in a Hilbert space with *unitary* metric) is thus not a mathematical trick that could in principle be avoided, as in [many parts of] classical physics, but an essential characteristic of the theory, to wit: a *necessary condition for obtaining meaningless answers to meaningless questions*.

Formula (5) covers two essentially different cases. If X and Y are *complementary* so that P_X and P_Y do not commute $\text{prob}_2(X; Y)$ is a *transition probability* which may take any *real* value between 0 and 1. If P_X and P_Y commute and hence X and Y are *commensurable* or *non-complementary* $\text{prob}_2(X; Y)$ is a *relative frequency* that takes only *rational* values between 0 and 1.

If these relative frequencies are written as fractions numerator and denominator are to be interpreted either as *statistical weights* of degenerated states or else as the frequency of a non-degenerate state in a certain *mixture* [non-uniform ensemble = ensemble of like systems in different states].

If only non-degenerate states ('pure cases') are considered, we have

$$Tr\, P_X = |(\psi_X, \psi_X)|^2 \, [= 1] \tag{6}$$

and

$$Tr\, (P_X P_Y) = |(\psi_Y, \psi_X)|^2 \tag{7}$$

where ψ_X, ψ_Y are the *state vectors*. If the latter are normalized, expression (6) takes the value 1 and (5) takes the familiar form

$$\text{prob}_2\,(X; Y) = |(\psi_Y, \psi_X)|^2. \tag{8}$$

The mathematical expressions on the right-hand sides of (5)–(8) are *invariants under the group of unitary transformations* and hence independent of the choice of the orthonormalized system of reference in Hilbert space. Different choices correspond to what are called different *representations* of the theory. The representation in which the position operator is diagonal, i.e., in which the eigenvectors of this operator are chosen as system of reference, is called the *Schroedinger representation*. In the *p*-representation often used by Dirac the momentum operator is diagonal, and in the *Heisenberg representation* the energy operator is diagonal. A further representation [used in perturbation theory] is the *interaction representation* in which the interaction energy is diagonal. The choice of a representation, just as the choice of a coordinate system in ordinary space, may be of practical importance for computations but has no fundamental import: by a unitary transformation any representation can be obtained from any other one.

5. The Quantum Mechanical Concept of 'Physical Quantity' and its Relation to Quantum Mechanical Properties (Modes of Reaction)

Historically, the concept of a quantum mechanical quantity or 'observable' resulted from a reinterpretation of the classical concept in the spirit of the correspondence principle. In classical mechanics the state of a system with f degrees of freedom is fixed by the values of $2f$ variables, to wit: f general coordinates q_i and f canonically conjugated momenta $p_i (i = 1, ..., f)$. Hence any continuous function of the q_i, p_i is taken to represent a physical quantity. In quantum mechanics the variables q_i, p_i are replaced by operators Q_i and P_i satisfying the *commutation rules*

$$[P_i, Q_k] =_{\text{df}} P_i Q_k - Q_k P_i = \pm \sqrt{-1}\, \hbar\, \delta_{ik}. \tag{1}$$

Hence only such operator-valued functions of the Q_i, P_i could be admitted as physical quantities that possess a real-valued eigenvalue spectrum. In the language of [meta]mathematics such quantities are called *hypermaximal Hermitean operators* (J. von Neumann). They are characterized by the fact that they admit of a representation in the form

$$A = \int\limits_{-\infty}^{+\infty} a\, dE^A(a) \tag{2}$$

where the integral is a Stieltjes integral and where the operator-valued function $E^A(a)$ is the so-called *resolution of unity* belonging to the operator A. Now these $E^A(a)$ are a *one-parameter family of commuting projection operators* with the properties

$$E^A(-\infty) = O, \quad E^A(+\infty) = I$$
$$E^A(a_1) E^A(a_2) = E^A(a_1) \quad \text{if} \quad a_1 \leqslant a_2.$$

(3)

(Where the function $E^A(a)$ is discontinuous we have discrete eigenvalues of A, where it is continuously increasing we have the continuous part of the eigenvalue spectrum of A.)

Thus, historically the projection operators entered quantum mechanics in a roundabout way *via* the concept of quantum mechanical quantity and its mathematical analysis.

In a rational reconstruction of the theory this way is to be reversed. We know already that properties of quantum mechanical systems are reactive properties, to be represented mathematically by projection operators. Hence, equation (2) is to be considered as the *definition of the quantum mechanical concept of physical quantity: any resolution of unity $E(a)$ defines, according to (2), a quantum mechanical quantity.* This definition is completely independent of all correspondence considerations and thus does not borrow anything from classical mechanics.

A correspondence to classical mechanics is only established by the following *first axiom of correspondence*: if, for a system with f degrees of freedom, we have $2f$ operators P_i, Q_i satisfying (1) these operators correspond to the general coordinates and momenta of classical mechanics.

A question that suggests itself in this connection is this: is there a classical quantity, i.e., a continuous function of the q_i, p_i, to every quantum mechanical quantity, i.e., to every resolution of unity? This question cannot be answered for sure at present. However, it can be shown mathematically and emerges most clearly from the quantum mechanical theory of diffraction that there is a *continuum* of quantum mechanical quantities between P and Q, all with a continuous eigenvalue spectrum.

The connection between quantum mechanical *quantities* and quantum mechanical *properties* is as follows. Let $E^A(a)$ be any resolution of unity defining a quantum mechanical quantity A according to (2). Then: the projection operator

$$E^A(a_1, a_2] \underset{\text{df}}{=} E^A(a_2) - E^A(a_1) \quad (a_1 < a_2)$$

represents the following property: *in case of a 'measurement of A' the quantity A takes a value from the interval* $(a_1, a_2]$. *A 'measurement of A'* is, by definition, any external action that forces the system considered to jump into one of the eigenstates of A.

6. THE DYNAMICAL LAW OF h-MECHANICS–BOHR'S RELATION OF INDETERMINACY

The dynamical law of h-mechanics determines the time dependence of the transition probabilities. It is a generalization of (3.5) and reads

$$\text{prob}_2(X; Y; t) = \frac{Tr\, P_X S(t)^{-1}\, P_Y S(t)}{Tr\, P_X}. \tag{1}$$

Here, $S(t)$ is the unitary transformation operator

$$S(t) = \exp\left(\frac{i}{\hbar} Ht\right), \quad H = \text{Hamilton operator}. \tag{2}$$

Mathematically, (1) may be interpretated in two [different but equivalent] ways, known as *Heisenberg picture* and *Schroedinger picture*, respectively. The Heisenberg picture rests on the mathematical identity

$$P_{U^{-1}YU} = U^{-1}P_Y U \tag{3}$$

for any unitary operator U. In view of this we have

$$S(t)^{-1}\, P_Y S(t) = P_{Y(t)}$$

with

$$Y(t) = S(t)^{-1}\, Y\, S(t). \tag{4}$$

Thus, in the Heisenberg picture the time dependence of the transition probabilities is attributed to a time dependence of the 'observables' [or rather: the reactive properties] while the state vector, here represented by the projection operator P_X, is considered time independent.

In the Schroedinger picture it is just the other way round. Since the trace is invariant under a unitary transformation of its argument we can write (1) in the form

$$\text{prob}_2(X; Y; t) = \frac{\text{Tr } S(t) P_X S(t)^{-1} P_Y}{\text{Tr } S(t) P_X S(t)^{-1}} \tag{1'}$$

The projection operator

$$P_X(t) =_{\text{df}} S(t) P_X S(t)^{-1} \tag{5}$$

corresponds to the time-dependent state vector

$$\psi(t) = S(t) \psi(o) = \exp\left(\frac{i}{\hbar} Ht\right) \psi(o) \tag{6}$$

which is the formal solution of the SCHROEDINGER equation

$$\frac{i}{\hbar} H\psi = \frac{\partial}{\partial t} \psi. \tag{6'}$$

Equation (2) or (6') may be called the *second correspondence axiom*. Together with (1) it says that the time dependence is uniquely determined by the Hamilton function, just as in classical mechanics. It should, however, be noted that (1) demands merely the existence of a time dependent unitary transformation and not the existence of a Hamiltonian. This has proved of great importance for the further development of quantum theory: The modern theory (S-matrix theory, dispersion theory) works only with a unitary S-operator which exists also in cases where a Hamiltonian H may not exist. It is certainly a further advantage of the present reconstruction of quantum theory that it gives logical priority to the S-operator over the Hamilton operator.

In the above equations the time variable t appears as a seemingly classical (non-quantized) quantity, in contrast to the coordinates [and all other quantities]. However, [the Schroedinger equation] (6') can be considered as a formal solution or representation of the operator equation

$$H T - T H = - \hbar/i \tag{7}$$

for energy and time, [corresponding to (5.1)]. In line with this we have an indeterminacy relation

$$\Delta E \, \Delta t \geqslant \hbar \tag{8}$$

between energy and time. However, [as time is not a state variable in either classical or quantum mechanics] the physical interpretation of (8) is somewhat different from that of the Heisenberg relation

$$\Delta p \ \Delta q \geqslant \hbar$$

[which of course follows from the probabilistic interpretation of the general formalism]. According to Bohr, (8) means that within a time interval Δt the energy of a system can only be determined up to $\pm \Delta E$ [with a reasonable degree of certainty] and [more important] that in a state of mean lifetime Δt the energy of the system is only determined up to $\pm \Delta E$. This interpretation has proved correct; it gives, e.g., the empirically known relation between the mean lifetime of excited atoms and the coherence length of the light emitted by them [on the one hand and the natural line breadth on the other hand].[2]

NOTES

* Translated from 'Grundlagen der modernen Physik – Teil III: h-Theorie (Quanten-mechanik)', in *Mikrokosmos-Makrokosmos*, Vol. **2** (ed. by H. Ley and R. Löther), Berlin 1967.

[1] [Viz., without meaning unless some meaning can be derived from the semantic axioms of the whole theory in an indirect way.]

[2] [This corrects a careless mistake in the original text. Coherence length Δl and mean lifetime Δt are of course classically related by $\Delta l = c \ \Delta t$. On the other hand, the natural line breadth in frequency measure is given by $\Delta v = \Delta E / h$. Together with (8) this yields $\Delta v \geqslant c / \Delta l$. Since this result can also be derived from classical Fourier analysis, an inconsistency would arise if (8) would not hold.]

CHAPTER XX

QUANTUM THEORY AND PHILOSOPHY*

INTRODUCTION

In his fundamental address inaugurating this Symposium Professor Harig has called attention to the task of further developing and enriching dialectical materialism by generalizing philosophically the novel results of natural science. The best way natural scientists can contribute towards this end is by giving a clear and correct picture of the pertinent theories.

In this respect many sins have been committed – as far as quantum theory is concerned – by physicists from either camp, Marxist or anti-Marxist. As you know, this has led to a remarkable confusion of philosophical fronts in the past years: objective idealists, mechanical materialists, and some Marxist philosophers and physicists could be seen to unite in a struggle *against* 'orthodoxy' and *for* a revision of modern physics in the spirit of mechanical determinism. If today this period of [philosophical] confusion is past we owe this mainly to the clarification effected by physicists such as W. A. Fock and A. D. Alexandrow whose enlightening work deserves special mention.

To be sure, I don't wish quantum theory to be excepted from philosophical criticism; a correct critique can only advance the development of physical theory. However, such a critique has to refer to the present theory *as it really is*, and not to some picture of the theory retouched philosophically or otherwise. The picture of quantum mechanics that was current and familiar did indeed show strongly positivist features; this could not be otherwise as the founders of that theory knew materialism only in its old-fashioned form, viz., mechanical determinism. While in years past the dominant trend was that of taking the positivistic distortion as the true picture of quantum mechanics and hence to criticize that theory as being idealistic, the opposite trend can be noticed today: quantum mechanics is now sometimes endowed with dialectical features beyond those actually present. This is the other mistake that has to be avoided, not only for truth's sake but also because a misjudgement of this kind would prevent any progressive criticism of the theory.

After these preliminary remarks I come to our theme proper. I shall try to extract and display those features of modern quantum theory which I consider as particularly important for a correct understanding and a correct philosophical judgement and evaluation of the theory. I shall concentrate mainly on features that have been somewhat neglected in previous discussions.

I

The first fact to be noted is that we are confronted not with one but with *two* quantum theories: (non-relativistic) *quantum mechanics* and (relativistic) *quantum field theory*. It would be nice if the former would be contained in the latter as a limiting case such as classical mechanics [or rather: its kinematic section][1] is contained as a limiting case in relativistic mechanics: there would be no need then to consider quantum mechanics separately. Unfortunately the relation between the two quantum theories is much more complicated. This may already be seen from the fact that there is nothing in quantum field theory that would correspond to the indeterminacy relations between position and momentum known from quantum mechanics. To understand why quantum mechanics cannot be obtained from relativistic quantum theory by a limiting process [or otherwise] two facts are decisive:

(i) A relativistic field theory does not possess a non-relativistic limit (except the trivial one where there is no motion at all). This results from the fact that any such theory rests on differential equations in which the differentiation with respect to time has the factor $1/c$ ($c =$ velocity of light in vacuum) so that in the limit $c \to \infty$ the differentiation with respect to time vanishes identically. This has nothing to do with [the specifics of] quantum theory.

(ii) A relativistic generalization of quantum mechanics would be a 'relativistic quantum mechanics' which for $h \to 0$ ($h =$ Planck's constant) goes over into classical relativistic mechanics. Such a theory is impossible for mathematical reasons as first shown by O. Klein for Dirac's theory and later by W. Pauli for the general case. In other words: *the union of relativity and quantum theory always leads to a quantum field theory.* (The reason for the non-existence of a 'relativistic quantum mechanics' is the well-known fact that the energy-momentum relation of relativistic point mechanics is non-linear but quadratic.)

On the other hand it would be wrong to say that there are no relations at all between quantum mechanics and quantum field theory – such a state of affairs would be possible only if the two theories were to refer to different objects. Excepting particles of zero rest mass (photons, neutrinos, and – possibly – gravitons) for which a non-relativistic theory cannot exist, quantum mechanics and quantum field theory do refer to the same objects. In some respects quantum field theory is indeed a generalization or [rather] 'Aufhebung' [dialectical negation] of quantum mechanics; e.g., while the number of particles is a constant of motion in quantum mechanics it is a dynamical variable, represented by an operator, in quantum field theory.

In view of these [latter] facts a [separate] discussion of quantum mechanics may appear out of date and redundant. Perhaps this is the right standpoint. The reason why this may be doubted lies in the fact that quantum mechanics is *not* a limiting case of [present] quantum field theory and that we have no reason to assume that this is merely a result of the deficiencies of present (local and linear) field theory.

II

Let us then begin with the features that are common to both quantum mechanics and quantum field theory, and hence characteristic of quantum theory as a whole.

(1) [*State description*]. The *state* of a physical system is represented not by the values of any classical quantities (such as coordinates and momenta, or field strength) but by a *direction* (*ray*) in an infinite-dimensional complex vector space (Hilbert space, HS); the unit vector in this direction is called *state vector*. This alone shows that the quantities known from classical physics have lost much of their significance in quantum physics.

(2) [*Change of state – I*]. If the system remains undisturbed, the direction of the state vector changes continuously in a well-defined way according to a so-called Schroedinger equation.

(3) [*Change of state – II*]. If the system is disturbed suddenly (viz., in

a non-quasistatic way) the state vector jumps into a new direction. Here, two cases must be distinguished:

(a) The disturbance has *not* the character of a measurement (examples: diffraction and scattering, i.e., disturbance by a screen with holes or by a potential, respectively); in such cases the change of state is 'deterministic' [i.e., uniquely determined by the disturbance] so that the state after the disturbance is uniquely fixed by the disturbance and the initial state; (the operator transforming the initial into the final state vector is a projection operator for pure diffraction and a unitary operator for scattering).

(b) The disturbance has the character of a *'measurement'*, i.e., it does not enforce a *definite* change of *state* but ensures that the state of the system becomes one of the eigenstates of a *definite quantity*. In this case the change of state is 'indeterministic' [i.e., not uniquely determined], but the *probability of transition* from initial state v_0 to any of the possible final states v_i is uniquely determined by the angle between the initial and the final state vector; (the state vectors v_i, $i = 1, 2, 3, ...$, correspond to the possible values of the 'measured' quantity – cf. 4).

(4) [*Physical quantities*]. To any direction, and hence to any state vector v, belongs a projection operator P_v. If the set $\{v_i\}$ is a *complete* set of orthogonal vectors in HS every operator of the form

$$A = \sum_i a_i P_{v_i}$$

with real numbers a_i represents a *physical quantity in the sense of quantum mechanics*, meaning: if the system is in state v_i the quantity A has the value a_i. If the system is in a state different from any of the states $\{v_i\}$ the quantity A has no definite value at all.

In the transfinite set of quantum theoretical quantities some correspond to the well-known quantities of classical theory: in this way a 'correspondence' to classical theory can be established. (Historically, this correspondence has played a great role: at first only such quantum theoretical quantities were considered to which there exist corresponding quantities in classical theory; this suffices for many [but not all] applications.)

(5) [*Concept of 'measurement'*]. Under (3b) a general definition of 'measurement' in the sense of quantum theory is given; (in current textbooks the concept remains undefined). This general definition does not

make any use of classical concepts. Hence the question arises as to its relation to the classical concept of measurement. Two remarks must suffice here.

(i) As far as *quasi-classical* quantities are concerned, i.e. quantities to which corresponding quantities exist in classical theory, the two concepts of measurement are effectively the same; this is to say that in either case the 'measured values' can be found by applying the classical theory of the measuring devices. Some caution however is required in interpreting a given experimental arrangement as a measuring device [for quantum theoretical quantities]. (In diffraction experiments, e.g., the Frauenhofer arrangement represents a device for measuring [quantum mechanical] momentum!)

(ii) Many [quantum mechanical] measurements end up with the particle [quanton] being absorbed; in such cases a 'final state' [of the quanton] does not exist, or rather: the final state ceases to exist the moment it comes into being. This somewhat paradoxical situation may be taken as evidence for conceptual deficiency of quantum mechanics: it does not occur in quantum field theory.

CONCLUSIONS

A. [*Causality problem*]. As pointed out under (3b), quantum theoretical probabilities refer to *transitions* from one state into another state under the action of some particular external agent; (the so-called *spontaneous* transitions in the case of light emission or radioactive decay are [likewise induced and hence] not genuine exceptions [as the theoretical treatment shows], but the details cannot be explained here). Hence quantum theoretical probabilities do not refer to statistical ensembles [as the distribution probabilities of statistical mechanics do]. It is, therefore, misleading to call quantum theory a statistical theory. In fact, the quantum theoretical calculus of [transition] probabilities does not coincide with the classical calculus of probabilities as formulated, e.g., by KOLMOGOROFF[2]. Hence the search for 'hidden parameters' is completely irrational: such parameters would have meaning only if the problem were that of explaining a statistical distribution. In contrast to the classical concept of probability, the quantum theoretical concept of transition probability is a new invention in the history of science, anticipated neither by mathematicians nor by philosophers; it represents [[a rational generalization and]] a dialectical negation of the deterministic concept of causality.

A striking characteristic of both the novelty and the irreducibility of quantum theoretical probability is the quantum theoretical state description. Though the state vector represents an objectively existing state, this state can be [semantically] defined only by the probabilities with which the systems reacts to certain external influences. Even if the state vector can be specified from a knowledge of the prehistory of the system (without measurement intervention), as is often the case, such a specification is – strictly speaking – always more or less hypothetical and can only be tested by the probabilistic reaction of the system to external influences. Obviously, this situation is a novel and interesting example of the dialectical relationship between *potential* and *actual*.

This conclusion is of course incompatible with the view – advocated here by Prof. Heber – that the probabilistic [stochastic] nature of quantum theory results from the [alleged] fact that in any measurement the quantity measured is always a characteristic of either the particle or the wave picture, and hence of a model that is never quite correct. First, such concepts as energy and momentum are not bound to any particular model. Second, the quantum theoretical concept of measurement as defined above is far more general than the classical one and not restricted to [quasi-]classical quantities. For instance, in diffraction experiments any arrangement with a *finite* distance between diffractor and photographic plate or scintillation screen is a measuring arrangement for a certain quantum mechanical quantity admitting neither corpuscular nor wave theoretical interpretation; in spite of this there exist only *probabilities* for the possible results of such a 'measurement'. Hence, I consider the named view as a holdover from the period when the new wine was filled into old bags [as the Germans express it], viz., when the classical concepts were still considered sufficient [or even indispensible] and the novel character of quantum theory was tried to be accounted for *solely* by using these concepts in a complementary way. In such a conception complementarity and probabilism were bound to appear as a restriction or even renunciation [of justified methodological demands] due to the use of inadequate concepts or inadequate experimental questions so that the opponents of quantum theory [such as Einstein] would be justified. Yet if quantum theory is considered as an (approximately) correct conceptual reflection of real properties of matter as required and vindicated by scientific practice, the conceptual apparatus of the theory has to be understood without reference to classical theory; only then will it become possible to give a correct characterisation of its relation to classical theory. Quantum theory will then appear as [[a generalization]] [a dialectical negation][3] but not as a restriction of classical theory.

B. [*Conceptual independence of quantum theory*]. As outlined above, all [general] quantum theoretical concepts, including the concepts of [quantum] physical *quantity* and *measurement*, can be reduced to the state concept without borrowing anything from classical theory. Thus, quantum theory is no less conceptually self-contained than other physical theories contrary to a widespread opinion that gives quantum theory an exceptional postition among physical theories [by maintaining that it cannot be formulated without the use of classical concepts]. As far as this opinion is more than a residue from the early stages of the theory it is based on the argument that the concepts of classical physics are not only indispensable in practice but also primary in epistemological respect because they are used to describe our (macroscopic) *experience*. Obviously, this argument is valid from the standpoint of positivism only.

Sometimes the alleged epistemological primacy of the classical concepts is founded on an appeal to *Anschauung* (visualizability) rather than to experience. Such a Kantian argument overlooks that human capacity for visualizing is capable of development, as rightly pointed out here by Professor Heber.

III

So far hardly anything has been said about the *wave-particle problem*. The reason is that quantum mechanics and quantum field theory give entirely different answers.

(6) Quantum mechanics is *not* a synthesis of wave and particle theory, though it does ascribe to the particles some properties that can be considered as a step towards such a synthesis:

(a) For a single particle with definite momentum the state vector in HS is a wave function representing a plane harmonic wave in physical 3-space. This however does not imply that the particle [quanton] always behaves as such a wave would do; it does so only as long as external influences [are absent or] do not have the character of a measurement. If they do have this character the probabilistic [stochastic] meaning of the state vector mainfests itself.

(b) If a system consists of several particles [quantons] interacting with one another then, even when the interaction has ceased to operate, a state vector cannot be ascribed to any one of the particles [quantons] but only to the whole system; the state vector is then [in the Schroedinger representation] a function of $3N$ coordinates if $N=$number of particles [quantons]. An analogous or corresponding situation exists only in classical point mechanics but not in classical field theory.

(c) If a system consists of *like* particles [quantons of the same kind] the state vector [in the Schroedinger representation] must be symmetric or antisymmetric in the particle [quanton] coordinates [i.e., its *commutational parity* must be $+1$ or -1], depending on the spin value, if correct results are to be derived. This is an additional postulate that does not follow from the general principles of quantum mechanics proper but does follow from those of quantum field theory [as first shown by Pauli]. The last remark applies not only – as generally known – to the [actual] *correlation* between spin and symmetry character [commutational parity], since even the *narrowing down* [of the field of possibilities left open by the general principles of quantum mechanics] *to the alternative* 'symmetrical or antisymmetrical' cannot be justified on the basis of the general axioms of quantum mechanics. The often advanced argument which refers to the 'indiscernibility' of like particles does not suffice for such a justification: if it did we would have to replace Boltzmann statistics by [the so-called][4] Bose-Einstein statistics already in the *classical* [statistical] gas theory. In fact, the restrictive postulate 'symmetric or antisymmetric' [for the state vector] expresses a much deeper and more fundamental property of particles [quantons] than mere indiscernibility, namely the fact that like particles [quantons] are the *quanta of a common field* and hence – contrary to like particles of classical physics – do *not* possess the properties of independent existence and genidentity. This makes it understandable that the named postulate follows from quantum field theory. Thus, this postulate represents a *partial anticipation of quantum field theory*.

(7) The quantum field theory can be considered as a synthesis, or rather as a *dialectical uplation*, of classical field and classical relativistic particle theory. As this seems generally admitted today, justification and explication [of this view] can be dispensed with.

(8) Quantum field theory has also done away with the old metaphysical

[undialectical] contradistinction between 'matter' and 'force': the quantized field is both the carrier of 'material' particles and (by way of the so-called 'virtual' exchange of such particles) carrier of force-like interactions. In contrast to this, quantum mechanics maintains the old contradistinction: force-like interactions are described as in Newtonian mechanics by potential energies, and the novel forces (so-called exchange forces) result from the additional postulate of symmetry or antisymmetry of the state vector and hence from a partial anticipation of quantum field theory as shown above.

CONCLUSIONS

C. Quantum field theory represents a tremendous advance over quantum mechanics, both from the physical and the philosophical point of view. Hence there can be no doubt that the further development has to start from quantum field theory and not from any attempt [past or future] at reinterpreting or modifying [the formalism of] quantum mechanics.

D. This conclusion is further supported by the fact that quantum mechanics is a completely closed theory without visible seeds of further development while quantum field theory does contain such seeds in the form of [certain] inconsistences, e.g., divergent expressions in case of interaction problems, and does offer various possibilities for generalization [or modification]. The study of these possibilities can be expected to lead to new physical insight that will be of considerable import to philosophy, too.

NOTES

* Translated from 'Quantentheorie und Philosophie', in *Naturwissenschaft und Philosophie* (ed. by G. Harig and J. Schleifstein), Berlin 1960.

[1] Cf. M. Strauss, 'Intertheory Relations', 1968 Salzburg Colloquium in Philosophy of Science; *Induction, Physics, Ethics*, (ed. by P. Weingartner and G. Cecha), Dordrecht, 1970.

[2] M. Strauss, 'Zur Begründung der Statistischen Transformationstheorie der Quantenphysik', *Sitz.-Ber. Berl. Akad. Wiss., Phys.-Math. Kl.* 27 (1936), 382–398.

[3] [Cf. note 1].

[4] What is called Bose-Einstein statistics has in fact been introduced into physics by M. Planck; cf. M. Strauss, 'Max Planck und die Entstehung der Quantentheorie' in *Forschen und Wirken*, Bd. I, Berlin 1960.

CHAPTER XXI

A SECOND FOUNDATION
FOR QUANTUM THEORY[1]*

As shown in a previous paper[2], the combination of complementarity logic and probability calculus is sufficient to establish the essential features of the general formalism of quantum theory, known as statistical transformation theory. The advantages of this way of founding quantum theory as compared to other procedures lie in the following:

(1) The complex-valuedness of the state vectors (unitary metric) is seen to be a direct consequence of complementarity logic.[3]

(2) The concept of *physical quantity* is reduced to the more general concept of *physical property*. Hence it can be *proved* that the operators representing physical quantities must be hypermaximal.

(3) The existence of a Hamiltonian is not required.

On the other hand, there are also some disadvantages:

(a) The use of the probability calculus (in its modified form as required by complementarity logic) blurs the fundamental distinction between *statistics* (probabilities of distribution) and *probabilistics* (probabilities of transition).

(b) The choice of the concept of *property* as primitive concept leads to preference being given to the Heisenberg picture as compared to the Schroedinger picture in the formulation of time dependence. Since in the Heisenberg picture two operators $A(t_1)$ and $A(t_2)$ do *not* in general commute so that the corresponding properties [or rather predicates] are inconnectible in the sense of complementarity logic, the necessity arises to give metalogical reasons for the inconnectibility of such predicates.[4]

Furthermore, the discussion that has taken place in the meantime has cast doubt on the necessity of using complementarity logic and has brought other fundamental questions into the foreground. An example of the latter is the question whether – or how far – quantum theory may be built up without the use of classical [or quasi-classical] concepts.

For these reasons the question of founding quantum theory has been investigated anew. A 'constructive' axiomatic system has been established, proceeding from general to more and more special axioms. The system is based on a generalized concept of *state* [as basic concept] which does not presuppose the existence of state determining or state depending quantities or properties.

The essential part of the system is a *classification of all possible types of state changes*. This allows one to *define* all essential concepts of quantum theory, including the concept of [quantum mechanical] measurement. No borrowing from classical physics is required. [[...]]

A more complete presentation of the new foundation outlined above will be given elsewhere.[5]

NOTES

* Translated from 'Eine zweite Begruendung der Quantentheorie', *Monatsber. Dtsch. Akad. Wiss. Berlin* **3** (1961), 532–534.

[1] The basic idea underlying the new foundation outlined in this paper has first been presented to the *Internationales Symposium anlaesslich der 550-Jahr Feier der* Karl-Marx-Universität Leipzig in October 1959; cf. M. Strauss, 'Quantentheorie und Philosophie' in *Naturwissenschaft und Philosophie* (ed. by G. Harig and J. Schleifstein), Berlin 1960 [Chapter XX this volume]. The first presentation *in extenso* was given to an ad-hoc Colloquium in the *Max-Planck-Institut für Physik und Astrophysik* in Munich, May 1961. The author is indebted to Professor W. Heisenberg [and his co-workers] for the hospitality extended to him.

[2] M. Strauss, 'Zur Begründung der statistischen Transformationstheorie der Quantenphysik', *Sitz.-Ber. Berl. Akad. Wiss., Phys.-Math. Kl.* **27** (1936), 382–398. [Chapter XVI of this volume].

[3] *Loc. cit.*, § 11.

[4] *Loc. cit.*, § 4.

[5] [Cf. Part III of 'Intertheory Relations', Chapter XXII of this volume].

INTERTHEORY RELATIONS (III)

QUANTUM MECHANICS AND CLASSICAL POINT MECHANICS*

1. *Restatement of QM*

(a) *Misrepresentations of QM: their common cause*

No other physical theory, including Einstein's General Theory, has been, and still is, more often and more thoroughly misrepresented than QM. Different causes have conspired to produce, not one or two, but a confusing wealth of such misrepresentations. Some of them originate in physical misinterpretations ('matter waves', 'pilot waves'), some in the vagaries and heuristics of the historical development ('duality'), some in bad philosophy ('uncertainty relations'), some in the normal difficulty of finding an appropriate new theoretical language for what cannot properly be expressed in the old one ('statistical interpretation'). Most of these early misinterpretations and misrepresentations have died out or have been corrected (transition probabilities instead of statistic distributions), but some still linger on, and, worse still, new ones have arisen and are still arising. On the semantic level we were offered 'causal interpretations' involving 'hidden parameters' and/or 'quantum potentials' – ghosts that never appear (except on paper), and, as the latest cry, 'Q-densities' – revived ghosts in a 'Ghost-Free Axiomatization'. On the axiomatic or semi-axiomatic level we were offered 'principles' or 'approaches' such as 'Principle of Indeterminacy', 'Principle of Superposition of States', 'Space-Time Approach', 'Differential-Space Theory', to mention but a few of them: to write the whole story of misinterpretations and misrepresentations would fill a book or two.

Is there any common ground to which the misinterpretations (and many of the remaining misrepresentations) can be traced back? I think there is, and I see it in the failure to realize that the operators representing so-called 'observables' refer not to ordinary but to dispositional properties or, more precisely, to *stochastic modes of reaction*, a stochastic mode of reaction being an *induced stochastic transition from one state to another one*, usually one of the eigenstates of some 'observable' which is then said to be 'measured'. But this latter term, borrowed from classical physics, is entirely misleading and merely gives rise to pseudo-problems and pseudo-theories such as the much-debated 'Quantum Theory of Measurement'. In fact, QM is a sort of *black box* theory in that the stochastic reactions (induced state transitions) are not explained (i.e., reduced to other processes) but taken as irreducible; what the theory does explain are the relative frequencies, or rather: the relative probabilities, with which the various possible reactions (induced state transitions) occur. Thus, *'reaction of a quantum system to macroscopic systems' is a built-in feature of QM* and, hence, a PI in terms of internal properties only is impossible or, at best, a metaphysical construct (with ghosts that never appear).

Furthermore, QM has no place for any preconceived Theory of Probability since the MF together with its probabilistic PI *logically implies* a mathematical theory of probability, and this theory is definitely *not* the classical theory of Reichenbach-Kolmogoroff[1]. However, it *is* in agreement with Popper's (*semantic*) propensity conception and may even *demand* it.

From what I have said it will appear not to be superfluous to explain the conceptual structure of QM in some greater detail by analysing typical applications. After all, the PI of any theory cannot be guessed from a mere inspection of its MF: it is through typical applications that the correct PI is brought to light, and this is particularly true of QM where the *feedback* from applications to the general theory has played a decisive role in finding the correct PI. The correct or *standard* PI will not be changed, i.e., there will be *no physical reinterpretation*[2]. But the standard PI has never been stated in adequate concepts – the nearest to such a statement being perhaps the axiomatic restatement of QM by Landé[3]; though it, too, is not entirely free from classical concepts (his 'particles'). Hence, a *reconceptualization*[4] of the standard PI *is* called for and will be given, at least in outline.

(b) *Analysis of applications: PI and conceptional structure of QM*

An analysis of typical applications of QM shows that most of them can be described in terms of *changes of state*, any *state* being represented mathematically by either a ray (direction) in an appropriate HS, or, by convention, by the *unit vector* $|\rangle$ in that direction, or, equivalently, by the *projection operator on that* direction $P_{a|\rangle} = P_{|\rangle}$, the relation between states, unit vectors $|\rangle$, and projection operators $P_{|\rangle}$ being one-one-one [5].

There are two essentially different classes of state changes: *stochastic* and *non-stochastic* ones, induced by *stochastic* and *non-stochastic* state changers, respectively.

A *non-stochastic* state changer, or rather its action on the state, is mathematically represented by either a *unitary* operator U or a *projection* operator P_1:

either

$$|\rangle_0 \to |\rangle_1 = U|\rangle_0, \quad P_{|\rangle 0} \to P_{|\rangle 1} = UP_{|\rangle 0} \tag{1}$$

or

$$|\rangle_0 \to |\rangle_1 = P_1 |\rangle_0, \quad P_{|\rangle 0} \to P_{|\rangle 1} = P_{P_1 |\rangle 0} \tag{2}$$

or by a combination of the two (such as PU or UP) – a case I shall not consider. Since these operators do *not* depend on the state $|\rangle_0$, they do in fact represent both the action of the state changer and the state changer itself.

State changers of P-type are the *diffractors* (diffractions screens etc.) and *they* are the only ones known.

For state changers of U-type we have two classes: (a) *time* and (b) *scatterers*. For case (a) we have

$$U\left(t_1, t_0\right) = U\left(t_1 - t_0\right) = e^{(i/\hbar) H(t_1 - t_0)}, \tag{3}$$

this being the solution of the Schrödinger equation for an arbitrary time-independent Hamiltonian H.

In case (b) the unitary operator U is known as S-matrix; hence we shall write U_S instead of U.

If all states $|\rangle_i$ transition to which can be induced by a *stochastic* state changer are othonormal:

$$_k\langle|\rangle_i = \delta_{ik}, \quad \mathrm{Tr}\, P_{|\rangle i} P_{|\rangle k} = \delta_{ik} \tag{4}$$

as is usually though not always the case, the stochastic state changer will be called either *stochastic separator* or *stochastic analyser*, depending on whether the 'induced' states persist after the process or not (they do not persist if the quanton gets absorbed in the process). If, in addition, the 'induced' states form a *complete* set $\{|\rangle_i\}$ in HS the analyser or separator is also called *complete*.

Now a complete set $\{|\rangle_i\}$ is *characteristic* not of *one* but of a *class of infinitely many* operator quantities all of the form

$$F \equiv F(\{a_i\}) \underset{\text{df}}{=} \sum a_i P_{|\rangle_i} \tag{5}$$

where the a_i are *arbitrary* real numbers, called the *eigenvalues* of F. Indeed

$$F |\rangle_i = a_i |\rangle_i. \tag{6}$$

Hence, if we speak of an *analyser* or *separator for some 'observable' A*, A always represents a *whole class* of 'observables' all having the same eigenvectors but different eigenvalues. This, by the way, is but one example of many to show that the concept of *state* is far more fundamental than the concept of *'observable'*.

The following block diagram gives a summary:

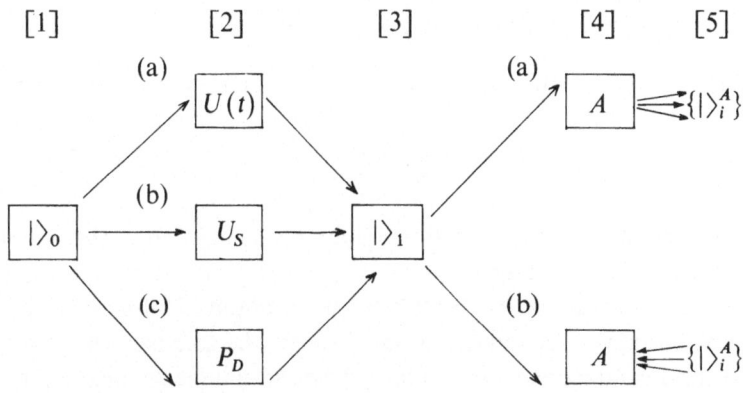

Initial state	Nonstochastic state changers:	Changed state:	Stochastic state changers:
	(a) time	$|\rangle_1 = U(t)|\rangle_0$	(a) Separator
	(b) scatterers	$|\rangle_1 = U_S|\rangle_0$	(b) Analyser for
	(c) diffractors	$|\rangle_1 = P_D|\rangle_0$	class A 'observables'

The probability for the induced transition $|\rangle_1 \rightarrow |\rangle_i^A$ is given by the theory:

$$\text{prob}_2 \left(|\rangle_1 \overset{A}{\underset{\Omega}{\rightarrow}} |\rangle_i^A\right) = \left|_0\langle|\Omega|\rangle_i^A\right|^2 = \text{Tr } P_{\Omega|\rangle_1} P_{|\rangle_i^A} \tag{7}$$

which can also be written as $\cos^2 \Theta_{1i}$ where Θ_{1i} is the 'angle' between the two vectors $|\rangle_1$ and $|\rangle_i^A$. Since $|\rangle_1 = \Omega|\rangle_0$ where Ω is any of the operators listed under [2] we have

$$\text{prob}_2 \left(|\rangle_0 \xrightarrow[\Omega]{A} |\rangle_i^A\right) = \left|_0 \langle|\Omega|\rangle_i^A\right|^2 = \text{Tr}\, P_{\Omega|\rangle_0} P_{|\rangle_i^A} \qquad (8)$$

(the normalizing factor for $\Omega = P_D$ is suppressed in the second expression).

It should be clearly understood that our block diagram represents a conceptual structure rather than an experimental set-up. Indeed, in the *case of diffraction* the diffractor is both the nonstochastic state changer and an integral part of the stochastic analyser for class A observables (the other part being the photographic plate or scintillation screen absorbing the quanton), and the class A 'observables' does not contain 'position' but 'observables' depending, in the simplest case, on the distance d between diffractor and absorber, in the form

$$A(d) = U(d) P_x U(d)^{-1}, \qquad (9)$$

P_x being the momentum operator and $U(d)$ a unitary operator satisfying

$$U(\infty) = I, \; U(0) = \int dq_x e^{ip_x q_x / \hbar}.$$

This dependence on d substantiates Bohr's warning that in the quantum mechanical analysis of any 'phenomena' (here: diffraction patterns) the *whole* arrangement has to be taken into account. Unfortunately, this warning has never been understood by those who criticize what they call the 'orthodox interpretation'. In fact, the quantum mechanical treatment of diffraction is a *test case* for a proper understanding of QM, and it is a sore reflection that it cannot be found in any textbook.

There is one further point that should be noted. What can be read off from the diffraction pattern are the (relative) transition probabilities (6) or (7). They are *the same for all class A 'observables'*. Hence it is misleading in this case to speak of a 'measurement' of a particular 'observable' since no *eigenvalues* of any particular 'observable' are involved. The identification of the A class by one of its members as in (8), (9) is only necessary to *predict* the transition probabilities and, hence, the diffraction pattern. The class itself is uniquely determined by the diffraction pattern, given the knowledge of $|\rangle_1$ (or $|\rangle_0$ and P_D) and d, so that it could in principle be inferred from the diffraction pattern *without any 'observable' being 'measured'*.

Finally, it should be pointed out that the stochastic character of the state transitions has nothing to do with the thermostatistical nature of the separator or analyser (as was once surmised): if it had, the diffraction pattern would depend on the temperature of the diffractor!

There remains the question: what does an 'observable' A represent? Well, from the 'observable'

$$A = \sum a_i P_{|\rangle_i} \tag{10}$$

we can construct *two classes* of observables:

$$A[\varphi] \underset{\text{df}}{=} \sum \varphi(a_i) P_{|\rangle_i} \tag{10a}$$

and

$$A[U] \underset{\text{df}}{=} \sum a_i P_{U|\rangle_i} = \sum a_i U P_{|\rangle_i} \tag{10b}$$

where φ is any mapping $a_i \to b_i$ and U is any unitary operator. The operators of the φ-class (previously called 'class A observables') all have the *same eigenvectors*, those of the U-class have the *same eigenvalues*. This is the clue to the answer: *an 'observable' A represents two different classes and hence two entirely different concepts*, each of them definable as the 'abstraction class' with respect to the equality relation concerned. The φ-class defines a set of *states* and hence represents a class of stochastic analysers and separators. The U-class defines a set of *eigenvalues* and hence represents the QM analog of a CPM physical variable or 'quantity'. The 'observable' A, being the only common member of the two classes, represents both.

Thus we have the following correspondence scheme

$$
\boxed{
\begin{array}{l}
\{|\rangle_i\} \leftrightarrow \{A[\varphi]\}_{\text{all } \varphi} \nwarrow \\[2em]
\hspace{4em} A \\[2em]
a \to \{a_i\} \leftrightarrow \{A[U]\}_{\text{all } U} \swarrow
\end{array}
}
\tag{10c}
$$

There is nothing in CPM that corresponds to $\{A[\varphi]\}_{\text{all }\varphi}$. The arrow on the left is one-sided because there may not be a classical variable a to any operator set $A[U]$.

I have left out all problems requiring perturbation theory for their solution, in particular all problems involving interaction between quantons or between a quantum system and the electromagnetic field. These problems do not give rise to any new fundamental view points for the physical interpretation of the formalism. There are just two phenomena I should like to comment on because they are often quoted as evidence for the existence of truly indeterministic, i.e., non-induced stochastic behaviour: radio-active decay and the so-called spontaneous emission of light. The radio-active decay can be explained either as a 'tunnel effect' (interaction with a macroscopic potential) or else as a kind of evaporation, depending on the model used for the nucleus; neither explanation warrents the talk of 'indeterministic behaviour' in the sense explained. The so-called spontaneous emission of light from excited atoms must, according to its quantum electrodynamic theory, be considered as a state transition induced by the 'physical vacuum' with its well-known zero-point fluctuations of the electromagnetic field strength.

2. *Relations between QM and CPM*

(c) *The two state descriptions*

The fundamental difference between CPM and QM is revealed by a comparison of the two state descriptions:

In CPM, the state of a S_f is always represented by *the values* of $2f$ *physical variables*, e.g. in the Hamilton or Poisson formulation by the values of any set of canonically conjugate variables q_i and $p_i (i = 1, ..., f)$, or, equivalently, by a *point* in a $2f$-dimensional 'phase space'.

In QM, the state of a σ_f is always represented by a ray or *direction a* $|\rangle$ in a HS, or (with the usual convention), by a *unit vector* $|\rangle$ in HS, i.e., a point on the unit 'sphere' in HS, or, equivalently by the projection operator $P_{a|\rangle} = P_{|\rangle}$. Thus there seems to be no correspondence but only contrast. In particular, since a vector in HS has ∞ components, the CPM number $2f$ seems to be replaced by ∞:

$$2f \to \infty$$

with no trace of f left.

This contrast, though mathematically correct, is deceiving. *In any 'representation' of MF the state vector $|\rangle$ of σ_a is a certain function of f parameter values.*

We therefore write

$$|\rangle^f = |\gamma^f\rangle = |\gamma_1, \gamma_2, \ldots, \gamma_f\rangle. \tag{11}$$

Thus the CPM number $2f$ *splits up* into two QM numbers f_1 and f_2

$$2f \Big\langle \begin{array}{l} f_1 = f \\[2mm] f_2 = \infty \end{array} \tag{12}$$

So much on the formal (mathematical) aspects.

On the semantic side, the following has to be said. The $2f$ parameter values determining the state of a S_f are values of *physical variables*. On the other hand, the f values γ_i are not necessarily eigenvalues of f 'observables' as in

$$|\rangle^3 = \delta\,(p_x - p_x^0)\,\delta\,(p_y - p_y^0)\,\delta\,(p_z - p_z^0) = |p_x^0, p_y^0, p_z^0\rangle\,;$$

instead, they may be *quantum numbers*. This is the case if $|\rangle^f$ is an energy eigenfunction belonging to a non-degenerate energy eigenvalue.

Thus, in state descriptions there is *no general correlation between the number $2f$ of independent classical variables* and *the number of independent 'observables'* that have to be 'measured' to fix the state. If we are lucky, a *single* (energy) 'measurement' may fix the state vector. However, in all cases like this *the cooperation of the theory is needed to fix the state:* we must know how the 'measured' energy value depends on the quantum numbers, i.e., we must know the function $E_i = g\,(n_i^1, \ldots, n_i^f)$.

To summarize, the splitting up of $2f$ into $f_1 = f$ and $f_2 = \infty$ is, formally, a splitting up of one number into two. If the *semantics* of these numbers is taken into account, it represents the *splitting up of one concept into two concepts*, or, putting it the other way round, a *conceptual degeneracy* on the side of CPM. Thus we have an ITR typical of the relation between T_1 and its generalization T_2.

The degenerate concept of CPM that splits up is *'independent components of state'*. It splits up into *'algebraically independent components of a state vector'* (of which there are $f_2 = \infty$) and *'physically independent parameters of a state vector'* (of which there are $f_1 = f$).

Both the decrease of $2f$ to $f_1 = f$ and the increase of $2f$ to $f_2 = \infty$ are of fundamental physical significance. That of $2f \rightarrow f_1 = f$ is well known and has often been commented on. The increase $2f \rightarrow f_2 = \infty$ (which does not fit into the usual correspondence scheme) usually remained uncommented. It is, in a sense, a compensation for the decrease $2f \rightarrow f_1 = f$: together they imply, in the light of the full PI, that the CPM *trajectories in phase space* are *replaced* by *transition probabilities between states* or (in the case of an undisturbed system) by a mere rotation of the state vector.

(d) *The 'laws of motion'*

There are two ways to compare the two 'laws of motion', i.e., the laws on time-dependence, in the two theories, according to whether the 'Schrödinger picture' or the 'Heisenberg picture' is used for QM. We start with the former.

In the 'Schrödinger picture' we have

$$|\rangle_t = e^{iH_{op}t/\hbar}|\rangle_0. \tag{13}$$

There is a corresponding equation for CPM, though it is not well-known. It reads

$$\begin{Bmatrix} q_k \\ p_k \end{Bmatrix}_t = e^{D_H t} \begin{Bmatrix} q_k \\ p_k \end{Bmatrix}_0 \tag{14}$$

with

$$D_H = \sum \left(\frac{\partial H}{\partial p_i} \frac{\partial}{\partial q_i} - \frac{\partial H}{\partial q_i} \frac{\partial}{\partial p_i} \right). \tag{15}$$

Thus, the correspondence relation for the 'Schrödinger picture' reads

$$\boxed{ \frac{i}{\hbar} H_{op} \leftrightarrow \sum \left(\frac{\partial H}{\partial p_i} \frac{\partial}{\partial q_i} - \frac{\partial H}{\partial q_i} \frac{\partial}{\partial p_i} \right) } \tag{16}$$

The *functional form* (14) for the classical equation of motion may appear artificial and *ad hoc*. In fact, it is nothing of the sort: it is a definition of *canonical time*, or rather an equivalent of that definition which reads

$$\Omega(\tau_2)\,\Omega(\tau_1) = \Omega(\tau_2 + \tau_1) \tag{17}$$

where $\Omega(\tau)$ is the *propagator* (operator of motion) *in state space* defined by

$$S(t + \tau) = \Omega(\tau) S(t) \tag{18}$$

$S(t)$ being the element in state space representing the state at time t. If (18) is interpreted to mean: there exists a propagator $\Omega(\tau)$ such that (18) holds, (18) is an implicit (partial) definition of 'state space'. The postulate (17) defines *time metric* (i.e., 'equality of time intervals'). This is best seen when the general solution of (17), equivalent to (17), is introduced:

$$\Omega(\tau) = e^{\kappa\tau} \tag{17'}$$

so that (18) reads

$$S(t + \tau) = e^{\kappa\tau} S(t). \tag{18'}$$

A change of *time metric*, i.e. a non-linear regauging of the *time scale*, would destroy the relation (17) and hence the propagator would not retain the form (17'). Thus *any propagator of the form* (17') *implies canonical time* (metric). It follows: *the time metric is the same in CPM and QM*.

In the 'Heisenberg picture' the time dependence is thrown from the state vectors on to the 'observables':

$$A_t = e^{(i/\hbar) H_{op} t} A_0 e^{-(i/\hbar) H_{op} t} \tag{19}$$

or, equivalently,

$$\frac{d}{dt} A = \frac{i}{\hbar} (H_{op} A - A H_{op}). \tag{20}$$

In CPM the time dependence of any physical variable $v = v(q_i, p_i)$ can be written in the form

$$\frac{d}{dt} v = [v, H]_{\text{Pois}} \tag{21}$$

with

$$[u, v]_{\text{Pois}} \overline{\overline{df}} \sum_{i=1}^{f} \left(\frac{\partial u}{\partial q_i} \frac{\partial v}{\partial p_i} - \frac{\partial u}{\partial p_i} \frac{\partial v}{\partial q_i} \right). \tag{22}$$

Thus, for the 'Heisenberg picture' the correspondence relation reads

$$[a, H]_{\text{Pois}} \leftrightarrow \frac{1}{i\hbar} (AH_{op} - H_{op}A). \tag{23}$$

Now, as discovered by Dirac, (23) is but one instance of a *general correspondence between Poisson brackets* and *'quantum brackets'*, the latter being defined by

$$[A, B]_{\text{QM}} \underset{\text{df}}{=} \frac{1}{i\hbar} (AB - BA), \tag{24}$$

another instant being

$$[q_r, p_s]_{\text{Pois}} = [Q_r, P_s]_{\text{QM}} = \delta_{rs}. \tag{25}$$

For this reason the 'Heisenberg picture' is often held to be more fundamental than the 'Schrödinger picture'. From our point of view we must expect that the opposite is true since state vectors turned out to be more fundamental than 'observables'. The actual situation is as follows. The [,]_{\text{Pois}} is defined for any pair of differentiable functions of the q_i, p_i, while the [,]_{\text{QM}} is defined for any pair of operators. The decisive question arising therefrom is this: *is there a one-one relation between the RD's of the two bracket-functions*, or, equivalently, *is there a one-one relation between the set of differentiable functions $g(p_i, q_i)$* on the one hand and *the set of all operators qualifying as 'observables'*, i.e., 'hypermaximal' Hermitean operators, on the other? If there is not, there is no isomorphism between the structures $\{[\ , \ ]_{\text{QM}}, \text{RD}_{\text{QM}}\}$ and $\{[\ , \ ]_{\text{Pois}}, \text{RD}_{\text{Pois}}\}$ but at the most between one of these structures and a *substructure* of the other. In this case, the double arrow in (23) would have to be replaced by a single arrow (in one *or* the other direction) and hence the 'Heisenberg picture' would prove to be *less* fundamental, than the 'Schrödinger picture', at least from the point of view of correspondence.

Now the question, which is a purely mathematical one, has not been finally settled, indeed it has hardly been noticed. However, my contention is that we must replace (23) by

$$\boxed{[a, H]_{\text{Pois}} \rightarrow \frac{1}{i\hbar} (AH_{op} - H_{op}A)} \tag{23'}$$

or more generally

$$[a, b]_{\text{Pois}} \leftrightarrow [A, B]_{\text{QM}}$$

by

$$[a, b]_{\text{Pois}} \rightarrow [A, B]_{\text{QM}}.$$

In other words, I contend that there are *more* 'observables' than differentiable functions $g(p_i, q_i)$, 'more' not necessarily in the *general* sense of set theory[7] but in a more specific sense to be explained. The simplest of my arguments is this:

Consider any classical variable $a = g(p_i, q_i)$ and assume that there does exist the corresponding operator $A = g_s(P_i, Q_i)$ where the suffix s means 'symmetrized'. Then, if A is written

$$A = \int \lambda \, dE^A(\lambda)$$

any operator

$$A[\varphi] \underset{\text{df}}{=} \int \varphi(\lambda) \, dE^A(\lambda) \tag{26}$$

with *arbitrary* function φ is also hypermaximal Hermitean and hence (representing) an 'observable'. Now if $A[\varphi]$ could be written in the form

$$A[\varphi] = \varphi(A) \tag{27}$$

$A[\varphi]$ would correspond to $\varphi(a)$. However, for *arbitrary* φ, (27) is *not* satisfied. Moreover, besides (27) we have to demand that the function

$$\chi[\varphi](q_i, p_i) \underset{\text{df}}{=} \varphi(g(p_i, q_i)) \tag{28}$$

belongs to the RD of [,]$_{\text{Pois}}$, i.e., is *differentiable* with respect to all p_i, q_i and this, of course, is likewise *not* the case for arbitrary φ.

Thus, there can hardly be any doubt that the structures considered are *not* isomorphic and, hence, that *no one-one relation can be established in the 'Heisenberg picture'*, but at the most the asymmetrical relation (23').

There are two further points that call for comment. As first shown by Dirac, the Poisson brackets and the quantum brackets satisfy the same set of algebraic equations (functional equations), namely

$$[X, Y] = - [Y, X]$$
$$[X, k] = 0 \quad (k = \text{const})$$
$$[X + Y, Z] = [X, Z] + [Y, Z] \tag{29}$$
$$[XY, Z] = [X, Z] Y + X [Y, Z]$$
$$[X, [Y, Z]] + [Y, [Z, X]] + [Z, [X, Y]] = 0 .$$

If we could establish the set (29) together with

$$[Q_r, P_s] = \delta_{rs}$$

$$[F, H] = \frac{\mathrm{d}}{\mathrm{d}t} F \tag{30}$$

on sound physical grounds, we would have a common basis of abstract axioms or 'implicit definitions' for the primitives [,], Q_r, P_r, H for both CPM and QM. By the addition of 'branching' axioms specific for CMP and QM, respectively, we could then obtain both CPM and QM as models$_2$ of the abstract system (29)–(30), in about the same way as the Galilei and the Lorentz transformation groups are obtainable as models$_2$ (representations in the sense of group theory) of the same abstract algebraic velocity group.

The second remark concerns the physical meaning of the covariance group of the system (29)–(30). The group itself is known as the *group of canonical transformations* or, in CPM, as group of contact transformations. The essential feature of these transformations is that it admits mixing of the p_r's and q_r's reminiscent of the mixing of space and time coordinates in c-theory while the Lagrange equations of CPM do not admit such mixing. Thus, the canonical transformations group is the *widest* covariance group known for any formulation of CPM, and it applies only to the Hamilton and Poisson formulations of CPM. In QM, the group of canonical transformations is a *proper subgroup* of the invariance group of the Hilbert space viz., the group of *all* unitary transformations (rotations). Now the physical meaning of the latter is well-known: all orthonormal coordinate systems in HS and hence all 'observables' with a *complete* set of eigenvectors $\{|\rangle_i\}$, are *on the same footing*. The choice of any one of them represents the choice of a class of analysers and separators for the $\{|\rangle_i\}$-class 'observables' in exactly the same sense in which the choice of a system of pseudo-Cartesian coordinates in Minkowski space represents the choice of a particular inertial frame. As there is no similar physical interpretation or justification in CPM for the group of canonical transformations, the Hamiltonian or Poisson formulation of CPM must be considered as a *partial formal anticipation* of QM.

From this point of view it is not surprising that other theories such as quantum field theories and GRT obstinately resist all attempts at a Hamiltonian formulation.

(e) *Galilei covariance*

Let q_i be the Cartesian position vector of the ith mass point. Then $p_i = m_i q_i$. The 'proper kinematic' Galilei transforms are:

$$q_i^* = q_i - vt$$
$$p_i^* = p_i - m_i v. \tag{31}$$

Since v, t, m_i are all constants with respect to $\partial/\partial q_r$, $\partial/\partial p_s$, it follows that

$$[q_i^*, p_k^*]_{Pois} = [q_i, p_k]_{Pois}. \tag{32}$$

Thus, in *CPM the 'proper kinematic' (and, in fact, also the full) Galilei transformation is a canonical transformation*; in other words: *the Hamiltonian and Poisson formulations of CPM are Galilei covariant*, as was to be expected.

Before explaining the mathematical situation in QM, let us ask whether we should *expect* the MF of QM to be Galilei covariant. I contend we should *not*, and this for two reasons.

First, the Galilei group implies conservation of the quantity

$$N = Pt - EX \tag{33}$$

where P = total momentum, E = energy, and X = coordinate vector of centre of mass. In QM, the operators E and X do not commute, hence N is not Hermitean and thus does not qualify as 'observable'. Of course, we can symmetrize N to

$$N_s = Pt - \tfrac{1}{2}[EX + XE] \tag{33s}$$

which would be Hermitean, but this seems to be a poor way out.

The second reason is this. If the MF of QM were Galilei covariant, the Galilei group would have to be represented (in the group) theoretical sense) by a subgroup of the unitary group in HS. Now we know the physical meaning of a unitary transformation in HS from the discussion above: it represents the transition from one complete set $\{|\rangle_i\}$ to another set $\{|\rangle_i\}$, and hence from one class of stochastic analysers to another one, e.g., in diffraction from Fraunhofer to Fresnel analysers; and such a transition has nothing to do with the transition from one frame to another frame, Obviously, any such analyser or separator, or rather its mathematical representation, would have to be Galilei transformed as well to satisfy the so-called Principle of Relativity (i.e., the principle of equivalence of inertial frames). If this is done the transition probabilities

$$\text{prob}_2 \left(|\rangle_0 \underset{\Omega}{\overset{A}{\to}} |\rangle_i^A\right) = \text{Tr}\, P_{|\rangle_0} P_{\Omega|\rangle_i A}$$

should, and do, turn out to be invariant.

But the mathematical condition for this to be the case is somewhat *weaker* than the condition that the transformations concerned form a *true* representation of the Galilei group: they may form a *projective* representation, and under such a representation group the Schrödinger equation is indeed covariant[8]. In other and perhaps more familiar words: *the space of state vectors* (MF$_s$) is *not* a *true* representation space for the Galilei group, but if we admit projective representations the *'equations of motion'* are covariant.

Thus, from the standpoint of group theory, the mathematical situation may not appear entirely satisfactory though it is in full agreement with the standard PI. In fact, we may take an entirely different attitude towards the problem of covariance, the same I recommend for thermostatics. In the latter theory, *containing walls* are implied which define a preferential frame. If transformation theory (Lorentz group) is applied, it is largely a matter of convention how to transform the state variables (such as temperature) as no transformation law follows from their definition[8a]. If *covariance* is required, it turns out that *the only covariant formulation is the trivial one*, i.e., that in which all state variables are treated as *invariants*.

The same attitude may be taken with respect to Galilei covariance of QM: instead of the implied walls we have the nonstochastic state changers and the implied stochastic analysers or separators defining the preferential frame. The only difference is that the analysers or separators *can* be mathematically represented in the MF while the walls can *not*. The *decisive* point, however, is the same for both theories: *equivalence of frames is not generally equivalent with covariance under frame transformations*; it is so only if the problem concerned does *not itself* define a preferential frame.

(f) *Limit relations*

The comparison between QM and CPM shows that there is *no limit relation between the MF_S*: Hilbert space remains Hilbert space for $h \to o$. In this respect the situation is not different from that concerning the MF_S of Newtonian and Einsteinian mechanics: the Minkowski invariant

$$(\Delta S)^2 = c^2 (\Delta t)^2 - (\Delta L)^2$$

does not split up into two invariants ($(\Delta t)^2$ and $(\Delta L)^2$) for $c \to \infty$ but becomes infinite. But here, the transition to CPM can be made on the next higher theory level, i.e., kinematics: the Galilei transformation and the Newtonian addition of velocities are *exact limits* of their counterparts for $c \to \infty$. They are also *asymptotic* limits for $v/c \ll 1$.

Now a characteristic feature of QM is this: *on no level of MF is CPM an exact or asymptotic limit of QM*. The nearest we can get to such a limit relation is the well-known WKB-approximation the CPM counterpart of which is the Hamilton-Jacobi equation. But, as we have seen, time is only *one* of the non-stochastic state changers, and even if there were no others there would still remain the stochastic state changers implied by the transition probabilities between states. It is the preoccupation with the problem of selfpropagation or 'motion' that has obscured the view on the far more important problems of reaction and interaction. As far as such problems are treated in CPM (e.g. collision problems, the CPM analog of QM scattering) they are treated by application of the conservation laws which are equivalent to the laws of motions. By way of contrast, the scattering operator U_S has little and the diffraction operator P_D has nothing at all to do with the selfpropagator $U(t)$, or, for the matter, with the action function S in the phase $e^{iS/h}$, and hence the much advertised WKB-approximation is of no fundamental import.

There remains the question where we can find limit relations *on the level of applications*.

A popular answer says that CPM is the limit of QM for large masses.

The argument refers to the Heisenberg relation in the form

$$\Delta \dot{q} \Delta q \geqslant \hbar/m,$$

but then $m \to \infty$ does not imply $\Delta \dot{q} \Delta q = 0$, let alone $\Delta \dot{q} = \Delta q = 0$. Besides, the correct quantum mechanical description of a macroscopic body is not given by making m large but by treating it as composite system of many quantons. Hence, if there is any hope of obtaining CPM from QM, it is by $N \to \infty$, where $N = $ number of quantons in the macroscopic system.

Now the quantum mechanical description of such a macroscopic systems involves two new features. First, we have to use state vectors with permutational parity ± 1 for like bosons and like fermions, respectively; this of course holds for any composite system. Secondly we have to use *statistical thermodynamics* since most properties of a macroscopic system depend on its temperature. Thus, *the only level of application where a limit relation between QM and CMP can be expected is that of statistical thermodynamics.*

Now for systems that can be treated by the Boltzmann-Planck method ($S = k \ln W$) (as opposed to the general Gibbs method) the difference between CPM and QM reduces to the difference between 'Boltzmann statistics' on the one side and 'Bose' or 'Fermi' 'statistics' on the other side, corresponding to permutational parities $+1$ and -1, respectively. Hence, the transition to classical theory may here be identified with the transition of Bose or Fermi statistics (or both, depending on the system considered) to Boltzmann statistics and such a limit relation does indeed exist for $T \to \infty$, the best known examples being the ideal gases, including the photon gas represented by the Planck law.

If we have to use the Gibbs method (virtual ensemble) we may either employ the 'sum over states' ('partition function') formulation, for which CPM and QM are rival theories, or else we may start straight away from the QM definition of entropy due to von Neumann:

$$S = - \operatorname{Tr}(\rho_T \ln \rho_T) \tag{34}$$

where ρ_T is the statistical operator [9]

$$\rho_T = \sum_i^{\infty} e^{-E_i/kT} P_{|\rangle_i} \tag{35}$$

representing the canonical ensemble. As the Boltzmann and the Gibbs theory can be considered as different interpretations or applications of the same MF[10], we shall again obtain, on the level of application to all well-defined problems, the asymptotic limit relation

$$\boxed{QM^{TS} \xrightarrow[(T \to \infty)]{} CPM^{TS}} \tag{36}$$

Two remarks remain to be made. First, in realistic interpretation $T \to \infty$ implies $N \to \infty$: with $T \to \infty$ the mean velocities approach c and QM has to be replaced by QFT where N is a dynamical variable, not a constant.

Second, we have *not* explained why CPM is such a good approximation to reality for all macroscopic bodies with velocity $v \ll c$. I don't know of any solution of this problem that is acceptable, but it seems to me that a solution would involve the following points. On the physical side, we should realize that *any application of CPM to macroscopic bodies involves a partial anticipation of QM:* without QM there would be no atoms or molecules and hence no macroscopic bodies. On the formal side, we cannot hope to obtain limit relations outside the level of application *unless CPM is reformulated* in an entirely different way. One of the possible ways would be the introduction of 'macroscopic' 'observables' which would 'almost' commute: transition probabilities would then all be nearly 1 or 0. To establish the connection with QM the 'macroscopic' 'observables' would have to be defined somehow (as ensemble averages?) in terms of the QM 'observables'. There exist various steps in this direction that are well-known but I don't think that the program has yet been carried through in an unobjectionable manner.

(g) *Concluding remarks*

From the present point of view QM, though a logically closed theory, is but a halfway house on the road to QFT. The decisive step was the discovery that a consistent combination of h- and c-theory within the general frame of QM, as attempted by Dirac with his relativistic theory of the electron, was impossible: the reinterpretation of this theory enforced by the well-known Klein paradox and known as 'hole theory' involved the first prediction of antiparticles and the treatment of N as a dynamical variable or field energy quantum number. The so-called 'second quantization' used in QM as an alternative to the orthodox method of treating many quanton systems, proved a formal anticipation of QFT with its use of creation and annihilation operation as solutions of the basic field commutator relations. The *ad hoc* postulates concerning the connection between spin value and permutational parity proved consequences of QFT. Thus there can be no doubt that QFT is 'dominant' over QM.

Furthermore, it fits better into the mathematical tradition as far as group theory is concerned: the space of the field operators *is* a space of *true* representations of the Lorentz-Poincaré group. In spite of all this QFT suffers from the defect that interaction between different fields or their quantons is not an inbuilt feature of the theory and, hence, that it is unable to explain the mass spectrum and the coupling constants. The next 'dominant' theory required, usually called 'theory of elementary particles', is still in the making.

More likely than not, it will contain a new universal constant l, as the Heisenberg nonlinear spinor theory does, and it will be interesting to see whether or not, and on what level of theory, there are limit relations to the present QFT for $l \to 0$. Meanwhile, logicians of science should study the ITRs between QFT and QM to attain a better understanding of both.

Turning to the pragmatic aspects of ITR study, I think that its merits for better understanding, better presentation and better teaching of physical theories need no explication.

Not all will agree that ITR theory may become a heuristic instrument for finding new physical theories. However, we can extend our studies to ITRs of the *second order*, viz., to *relations between relations*. We may have ground for believing that the new theory (T_4) looked for will stand in the same (or a similar) relation to T_3 as T_2 stands to T_1:

$$T_4 : T_3 \simeq T_2 : T_1 .$$

In fact, this was precisely the heuristic scheme by which Schrödinger obtained his 'wave equation':

'wave mechanics': CPM $\simeq$ wave optics: geometrical optics.

From the standpoint of QFT this 2nd order relation is essentially correct: both the Maxwell field and the particle fields have to be quantized.

ITR study is, in a sense, the logical component in the History of Physics. This implies a second heuristic aspect: knowledge of the developmental laws of physics[11] is a help, if not a precondition, for a sound strategy of physical research.

NOTES

* Reprinted from *Induction, Physics, and Ethics* (ed. by P. Weingartner and G. Zecha), D. Reidel Publ. Co., Dordrecht-Holland, 1970.

[1] Cf. ref. 4 Part I. [Chapter XII].

[2] In fact, any PI different from the standard one would give a different physical theory. See also note 4.

[3] Cf. note 8, Part I. [Chapter XII].

[4] Many interpretations, offered as mere reconceptualizations, are in fact non-standard PIs. A typical example is Bohm's 'causal interpretation'. In spite of this author's contention to the contrary, his 'interpretation' gives entirely different results. For instance, it predicts that a hydrogen atom has a magnetic moment even in the ground state, in contradiction to the standard PI.

[5] This relation can be established in infinitely many ways, each one characterized by a phase factor exp(ia) *common to all* state vectors. This fact is often described by saying that the state vector is only defined 'up to an arbitrary phase factor'. This is utterly misleading as it suggests a one-many relation between states and state vectors.

[6] We omit the normalizing factor $(\operatorname{Tr} P_1)^{-1}$.

[7] The set theoretical distinction between cardinal numbers is the coarsest one possible. If we omit one $|\rangle_k$ from a complete set $\{|\rangle_i\}$ the resulting set $\{|\rangle_i\}_{i \neq k}$ has the same set theoretical cardinal number as the complete set though it is no longer complete. Thus, in HS we can distinguish between ∞ (meaning the cardinal number of a complete set) and $\infty - 1$.

[8] For detailed discussion and proofs cf. J.-M. Lévy-Leblond, 'Galilei Group and Nonrelativistic Quantum Mechanics', *J. Math. Phys.* **4** (1963) 776–88.

[8a] This view is now shared by most competent students of the question; cf. the last section 'Relativistic Theory' in *Procedings of the International Conference on Statistical Mechanics* (Kyoto, 1968) (Supplement to *Journal of the Physical Society of Japan*, Vol. **26**, 1969).

[9] A great deal of confusion has been generated by the use of statistical terms for which the referent, i.e. the ensemble, is not specified. Thus, the expression

$$\bar{A} \underset{\text{df}}{=} \langle |A| \rangle = \operatorname{Tr} P_{|\rangle} A = \sum |\langle |\rangle_i|^2 a_i \tag{a}$$

is usually spoken of as 'mean value' or – worse still – as 'expectation value' of A for (or in) the state $|\rangle$. This is *nonsense* if taken verbally since neither a mean value nor an expectation value is defined for a single system; it is *wrong* if the ensemble to which the mean value refers is taken to be the *uniform ensemble* ('pure case') of systems all in the same state: in this ensemble A has no mean value at all since none of the ensemble members is in a state where A has any definite value. In fact, the expression (a) is the *mean value* of (the eigenvalues of) A in the *nonuniform ensemble* ('mixture') in which the eigenstate $|\rangle_i$ of A occurs with the *relative frequency*

$$h_i = |\langle |\rangle_i|^2 = \operatorname{prob}_2 (|\rangle \xrightarrow{A} |\rangle_i) \tag{b}$$

To call expression (a) the 'expectation value of A' for a 'measurement' of A is also wrong since in general the value of expression (a) will be different from any eigenvalue of A and hence the 'expectation' of finding value (a) by a 'measurement' of A will be exactly zero!

[10] Cf. E. Schrödinger, *Statistical Thermodynamics*, Cambridge 1946.

[11] Cf. M. Strauss, 'Entwicklungsgesetze und Perspektiven der Physik', *Monatsber. Dtsch. Akad. Wiss. Berlin* **9** (1967) 538, 47: [Chapter II of this volume].

PART D

IN MEMORIAM HANS REICHENBACH

CHAPTER XXIII

HANS REICHENBACH AND THE BERLIN SCHOOL*

I

My acquaintance with Hans Reichenbach dates from the day of his inaugural lecture at Berlin University at the beginning of the winter term 1926–27. To the scientific world Reichenbach was then known merely as the author of 'Axiomatik der relativistischen Raum-Zeit-Lehre'[1] – a few earlier papers on the theory of probability were hardly known beyond a narrow circle of philsophers. Before being appointed professor at Berlin University he had been assistant to E. Regener and lecturer (Privatdozent) at Stuttgart [Technical College]. His Berlin appointment had been supported above all by von Laue, Planck, and Koehler, but also by Nernst, E. Schmidt, and von Mises, while Bieberbach, Diels, and Solger opposed the appointment.[2] The reason for this opposition appears to have been political. The issue emerges with sufficient clarity from a letter by E. Regener to Planck, dated 18.2.1925. In this letter Regener speaks of 'unexpected difficulties' and regrets that he cannot give a definite "Aufklärung über das, was Herrn Reichenbach aus Göttingen nachgesagt wird". The letter continues:

Was Reichenbachs allgemeine politische Richtung betrifft, so ist mir bekannt, dass er durchaus pazifistisch eingestellt ist. ... Nicht aber möchte ich glauben, dass er öffentlich für die Verweigerung des Heeresdienstes eingetreten sei. ...

For better understanding I recall the fact that Reichenbach, after having studied one year each at the universities of Berlin, Munich, and Berlin, had chosen the University of Göttingen for his fourth year of study. His subjects had been philosophy, mathematics, physics, and pedagogics. The records do not name the persons responsible for the political denunciation. To the honour of Göttingen University it can be said that D. Hilbert had sent a positive testimonial on Reichenbach to Berlin University.

Reichenbach's inaugural lecture had the title 'Kant und die gegenwärtige Physik'; it contained a clear refusal of Kant's *a priori*. This is remarkable in view of the fact that his doctor thesis 'Der Begriff der Wahrscheinlichkeit und seine Bedeutung für die mathematische Darstellung der Wirklichkeit' (1915) still rested on Kant's 'Kritik der reinen Vernunft', though it emphasized the materialist components of this work.

In the following six years (1926–1932) Reichenbach developed an extensive and intensive activity as a lecturer and researcher, which – together with numerous publications – led to what may be called the 'Berlin School'. Its organisational form was the 'Gesellschaft für empirische Philosophie' the managing committee of which also included, if I remember correctly, Wolfgang Koehler and Walter Dubislav in addition to Reichenbach.

To the Reichenbach group belonged, besides his students, above all K. Grelling and W. Dubislav; to the latter we owe the first comprehensive monograph on the problems of definition. Later V. Bargmann joined the group and is now well known among experts as a specialist in General Relativity.

In his lectures Reichenbach often developed new ideas which were thoroughly discussed in his seminars. The discussion was very frank and informal, thanks to Reichenbach's provocative support. Its usefulness has been acknowledged by Reichenbach himself; in the Preface to his 'Wahrscheinlichkeitslehre', probably his most important work, he writes:

Von größtem Nutzen war es für mich, daß ich die Gedanken des vorliegenden Werkes wiederholt, seit dem Jahre 1927, in Vorlesungen vor meinem Hörerkreis an der Berliner Universität darlegen konnte; in gemeinsamer Diskussion sind damals nicht nur viele Einzelheiten und Beispiele entstanden, sondern es wurde damit vor allem auch die Atmosphäre geschaffen, in der ich den mühevollen Weg zur Aufdeckung der Grundgedanken gehen konnte.[3]

Thus the much advertised unity of teaching and research was here living reality.

Reichenbach's best and most important publications fall likewise into his Berlin period:

Philosophie der Raum-Zeit-Lehre (1928)

Ziele und Wege der physikalischen Erkenntnis (1929) (Hdb. Phys.)

Ziele und Wege der heutigen Naturphilosophie (1931)

Axiomatik der Wahrscheinlichkeitsrechnung (1932)

Wahrscheinlichkeitslogik (1932).

To this must be added the preparatory work on

Wahrscheinlichkeitslehre (Leiden, 1935).

In order to escape the impending 'hell' (by which he meant the Hitler regime) Reichenbach accepted in 1932 a call by Istanbul University to become Professor of General Philosophy. From there followed an appointment at the University of California at Los Angeles where he wrote among other things

Experience and Prediction (Chicago, 1938)

and the controversial

Philosophic Foundations of Quantum Mechanics (Berkeley–Los Angeles, 1944).

He died there in 1953, aged 61.

A bibliography[4] of his work, published in 1959, lists 195 titles, among them translations into all the major world languages (excepting Russian) and also into Japanese and Hungarian.

A considerable part of these titles refers to articles in newspapers or periodicals or to booklets written for the general reader. In this Reichenbach follows in the tradition of Boltzmann, Haeckel, and other materialist propagators – with the difference that, according to the new situation, the polemic is directed against agnosticism and relativism in epistemology. A further reason for Reichenbach's activity as a writer [as far as his Berlin period is concerned] may have been his position as 'a.o.' [extraordinary] professor which called for additional income. Indeed, in the document of appointment, dated August 11, 1926, and signed by the Prussian Minister of Science, Art, and Education, we read: "Dagegen erwerben Sie keinen Ansprach an den Staat, insbesondere nicht auf Übertragung eines planmässigen Lehrstuhls". To be sure, the additional income could have been obtained by Reichenbach in a more convenient way by working for industry 'where he still has good chances because he has done very good work in the field of radiotelegraphy', as Regener testified in his letter to Planck mentioned above. That Reichenbach choose the less convenient way signifies that he was well aware of the social function of philosophy and tried to use it for promoting social progress.

II

In comparison with the writings of Carnap and the Vienna Circle, those of Reichenbach and the Berlin School have attracted little attention from Marxist philosophers. One of the main reasons appears to be that the scientific ingredient in the writings of the Berlin School is much stronger than in those of the Vienna Circle so that they are less easy to understand by philosophers lacking a physico-mathematical training. If the Berlin School is mentioned at all it is [mostly] treated as a sort of twin, or even a branch, of the Vienna Circle. Thus, in a book that appeared here as late as 1959 Reichenbach is called 'a former member of the Vienna Circle' ('ehemaliger Angehöriger des Wiener Kreises'). The mistake may partly be due to the fact that Reichenbach and Carnap were co-editors of *Erkenntnis*, the periodical that succeeded *Annalen der Philosophie*, and further to the fact that Reichenbach, in his public activities, exercised great restraint in voicing any opposition to the Vienna Circle.

The facts just mentioned were tactical measures, calculated to avoid a two-front struggle at least in professional life. To a man intent on promoting and asserting a philosophy in close contact with modern science the 'academic' philosophies of neo-Kantianism, phenomenalism, and 'Als ob', dominating the German university cathedrals at that time, were bound to appear as the main enemies. Hence a temporary alliance with Carnap as the leading head of the Vienna Circle was to be welcomed for practical reasons. That this interpretation is essentially correct can easily be proved by quotations from his works and letters which contain many attacks on the Viennese positivism. In fact, during his Berlin period Reichenbach was a conscious, though not always a consistent, materialist. In order to express his disapproval of logical positivism he called his own philosophical approach 'kritischer Realismus', 'logischer Empirismus', or 'wissenschaftliche Naturphilosophie'. Thus, he even used the disreputable term 'Naturphilosophie' instead of the modern 'Philosophie der Naturwissenschaften' because the latter had become almost a synonym for positivism.

In contrast to his pragmatic behaviour as co-editor, Reichenbach has made no conscious concessions to the positivism of the Vienna Circle in his writings, as a careful examination of his works shows. Already in 1921 he attacks Petzold, a disciple of Mach, for his positivist misinterpretation of Einstein's [so-called] Lorentz contraction:

Die Frage, ob die Lorentz-Kontraktion wirklich oder scheinbar ist, ist für Petzold eine leere Frage; für ihn bedeutet 'wirklich' dasselbe wie 'beobachtbar', und demgemäß kann etwas für einen Beobachter wirklich sein, was für einen anderen Beobachter nicht wirklich ist. An diesem Punkt weicht der Petzoldsche Posivitismus von der Relativitätstheorie ab. Einstein behauptet nicht die Relativität der Wahrheit, und nur eine einseitig positivistische Interpretation kann eine solche Behauptung in die Theorie hineinlesen.[5]

In a letter, dated March 30, 1936, to the present author Reichenbach replies to a remark concerning Carnap's *Logische Syntax der Sprache* as follows:

Sie haben ganz recht, es ist ein Schritt weg vom Dogmatismus der Wiener Schule. Besonders in seiner Tendenz gegen Wittgenstein ... Schlimm ist aber noch der Carnap'sche Konventionalismus. Der Begriff der Wahrheit geht bei seiner Auffassung ganz verloren ... Carnap hat bemerkt, daß solche Sätze (d.h. Sätze über die Außenwelt) nie als absolut wahr bewiesen werden können, und kommt nun auf die Idee, daß sie willkürlich sind ... Der Positivismus ist unhaltbar; er beruht auf der Verifizierbarkeitsforderung.[6]

Two years later, in a letter dated March 13, 1938, he criticizes Carnap's conventionalism in questions of syntax as follows:

Wir sind doch keine Schachspieler, die irgend ein neues Spiel erfinden wollen, für das man Vorschläge machen könnte; sondern die Wissenschaft hat ganz bestimmte Zwecke, und es ist die Frage, ob die betr. Sprache hierfür ein geeignetes Mittel darstellt. Wenn die Wissenschaft Voraussagen machen will, so ist es nicht willkürlich, welche Syntax man benutzt.[7]

With social conditions [in Germany] becoming more and more fascist, Reichenbach's [political] attitude became more radical. Thus, he sent his children to the Communist-led Karl-Marx-Schule in [the far-off district] Neukoelln, although he was not an adherent of Communism. Obviously his ideas of progressive education appeared to him more nearly realized by that school than by bourgeois schools. His interest in progressive education he had expressed long before. In the years 1912–14 no less than nine articles by him appeared in various periodicals, among them Ostwald's *Das monistische Jahrhundert, Die Freie Schulgemeinde*, and *Arbeiten des Bundes für Schulreform*; all of them were directed against the influence of the Church and the military powers on child education. Thus it was hardly a coincidence that he had married a woman who herself was a progressive pedagogue.

III

Considering the wide scope of Reichenbach's work it is of course impossible here to give a probative critical survey. Some hints must suffice.

It is advisable to distinguish between the programmatic writings of Reichenbach and those that are devoted to the logical analysis or axiomatic foundation of particular theories. The latter concern primarily the scientist, the former the philosopher.

We start with a central and still topical question from the philosophical field: the *relation between philosophy and natural science*. On this controversial topic Reichenbach had this to say:

Wer die Philosophie unserer Zeit mit der Arbeitsweise der großen philosophischen Systematiker des 17. und 18. Jahrhunderts vergleicht, dem tritt als grundlegender Unterschied die Verschiedenheit in ihrer Einstellung zur Naturwissenschaft entgegen. Während jene klassischen Philosophen im engsten Zusammenhang mit der Naturerkenntnis ihrer Zeit standen, ja z.T. selbst, wie *Descartes* und *Leibniz*, führende Mathematiker und Physiker waren, ist in unserer Zeit zwischen Philosophie und Naturwissenschaft eine Entfremdung eingetreten, die zu einer unfruchtbaren Spannung zwischen beiden Gruppen geführt hat. Die Philosophen, deren fachwissenschaftliche Schulung sich zumeist auf historisch-philologischem Boden vollzog, werfen dem Naturwissenschaftler zu weitgehende Spezialisierung vor und wenden sich geisteswissenschaftlichen Problemen zu; die Naturwissenschaftler andererseits vermissen in der Philosophie die Behandlung der erkenntnistheoretischen Probleme, die wohl von einem *Leibniz* oder *Kant* im Rahmen der damaligen Naturwissenschaft gelöst wurden, im Rahmen der heutigen Naturerkenntnis aber nach neuer Durcharbeitung verlangen. Eine gegenseitige Geringschätzung, die den Sinn der Denkrichtung des andern verkennt, ist der Ausdruck dieser inneren Trennung.

Blickt man historisch zurück, so kann man die Wurzeln dieser Spaltung durch das vergangene Jahrhundert hindurch verfolgen. Noch für *Kant* bildete der Erkenntnisbegriff der mathematischen Naturwissenschaft den Ausgangspunkt aller philosophischen Erkenntnistheorie; und wenn man darin auch mit Recht eine gewisse Einseitigkeit seines Systems begründet sieht, so liegt darin doch zugleich die Stärke seiner erkenntnistheoretischen Position, der seine Philosophie ihre große Auswirkung verdankt ...

Aber *Kants* Lösung des Erkenntnisproblems war zugleich auch die letzte, in der die Naturwissenschaft zum Ausdruck kam ...

Der Philosoph lebte der Meinung, daß die Probleme der Naturerkenntnis seit *Kant* gelöst seien, daß es sich für die Entwicklung der Naturwissenschaft nur noch um die Ausfüllung des Kantschen Programms handeln könnte – eine Auffassung, die auch in der biegsameren Form der neukantischen Schule vor Widerspruch mit der naturwissenschaftlichen Entwicklung nicht zu schützen war.

Es waren vielmehr die Naturwissenschaftler selbst, die im Laufe des letzten Jahrhunderts die Theorie der Naturerkenntnis zugleich mit ihrem inhaltlichen Ausbau ausgebildet haben. – So stehen wir vor dem eigent(üm)lichen Resultat, daß die Entwicklung der exakten Erkenntnistheorie im letzten Jahrhundert nicht von den Philosophen, sondern von den Naturwissenschaftlern vollzogen wurde, daß da, wo man auf einzelwissenschaftliche Dinge zielte, Erkenntnistheorie in sehr viel höherem Maße produziert wurde als da, wo man sie in philosophischen Spekulationen suchte.

Freilich ist die Situation allmählich auch für den Naturwissenschaftler zu kompliziert geworden. Auch er kann die eigentlich philosophische Auswertung nicht mehr vollziehen, einfach deshalb nicht, weil das einzelne Gehirn zur gleichzeitigen einzelwissenschaftlichen und philosophischen Arbeit nicht ausreicht. Eine Teilung der Arbeit ist unerläßlich geworden, seitdem sowohl die positive wie auch die erkenntnistheoretische Forschung eine solche Fülle von Kleinarbeit verlangt, daß sie die Kräfte des einzelnen übersteigt. Es kommt hinzu, daß philosophische und fachwissenschaftliche Arbeitsrichtung, so sehr sie in ihren großen Zügen aufeinander angewiesen sind, innerhalb der Mentalität des Einzelforschers doch geradezu entgegenwirken; die philosophische Besinnung auf Sinn und Bedeutung der Erkenntnis kann den Prozeß des naturwissenschaftlichen Erkennens geradezu hemmen, kann die Aktivität lähmen, die ohne eine gewisse Verantwortungslosigkeit den Mut zur Beschreitung neuer Wege nicht aufbrächte ...

Die Durchführung einer solchen Philosophie der Naturerkenntnis muß deshalb selbst einer besonderen Gruppe von Einzelforschern vorbehalten bleiben, wie sie sich in letzter Zeit deutlich herauszuheben beginnt; einer Gruppe, die einerseits die Technik der mathematischen Naturwissenschaft beherrscht, andererseits aber von ihr nicht derart belastet ist, daß sie über der Einzelarbeit den philosophischen Blick verliert.[8]

Worth emphasizing are the following points:

(1) In contrast to logical positivism, the taks of philosophy is not reduced to the mere analysis of real science but is conceived as a creative activity. A detailed reasoning in defense of this view can be found in Reichenbach's *Ziele und Wege der heutigen Naturphilosophie* (1931) where he attacks the positivistic conception of 'philosophy of natural science'. The theories of natural science should, according to his view, be considered merely as raw material for establishing a scientific natural philosophy, just as observations and experiments serve as raw material for establishing physical theories. In either case the creative process consists in the theoretical treatment of the raw material, viz., in the specific nature of the questions asked and in the scientific abstraction required to answer them.

(2) The second point worth emphasizing is his undogmatic approach. The relation between philosophy and natural science is not conceived as something absolute but as a historical category.

(3) From an analysis of the given historical situation Reichenbach finally draws the practical conclusion that a division of labour between natural scientists and philosophers trained in natural science has become unavoidable.

This last point calls for comment. Reichenbach speaks here only of 'philosophische Auswertung' [which in the given context means 'philosophic exploitation' rather than 'philosophic evaluation'] of given theories, having in mind no doubt relativity theory and quantum mechanics. A heuristic function of philosophy in gaining new theories is denied with [doubtful] psychological arguments. However, Reichenbach does recognize that *in the long run* advance in science and advance in philosophy depend on each other.

IV

A few principal remarks only can be made here on Reichenbach's presentation of the foundations of relativity theory and quantum mechanics.

The immediate import of *Axiomatik der relativistischen Raum-Zeit-Lehre*[9] [for that time] I see in the fact that it facilitated a better understanding of Einstein's ideas by explaining them in the form of a complete system of 19 axioms and 21 definitions. The complexity of this system results from its 'constructive' character aiming at a clear-cut separation between purely synthetic statements and conventional definitions. To this end only 'elementary facts' are admitted as axioms. By 'elementary facts' Reichenbach means objective laws that can be formulated without using arbitrary definitions and be tested separately with a minimum of theoretical assumptions involved. This idea of constructive axiomatics is a decisive step in the development of physical axiomatics. To be sure, it is adapted to the demands of logical analysis of a given theory rather than to the heuristic demands of constructing a new theory: in the latter case general principles are more useful than elementary facts.

In spite of this restriction constructive axiomatics as developed by Reichenbach will keep its value. The two forms of physical axiomatics [viz., the deductive and the constructive form] simply correspond to different aims [or rather to different approaches, as both forms aim at establishing the mathematical formalism together with its physical interpretation, in contrast to a third form of physical axiomatics which takes the mathematical formalism for granted and axiomatizes merely its physical interpretation]; hence the constructive and the deductive form of physical axiomatics are complementary to one another. Of course there exist many intermediate forms. It depends entirely on one's intention which form is chosen as most suitable. In general it can only be said that the axiomatic presentation of a physical theory constitutes a test of the soundness of the underlying conception. Reichenbach's axiomatics of relativistic space-time theory shows that he has stood this test.

V

The situation with respect to Reichenbach's *Philosophic Foundations of Quantum Mechanics* (1944)[10] is essentially different. The first thing to notice is that Reichenbach does not advance any axiomatic reformulation of the theory. Neither do we find a logical analysis that would disclose the philosophical basis [or the philosophical implications] of the theory such as the recognition of *chance* as an objective category, a thing done later by Fock. Instead, various 'interpretations' are being discussed that are said to differ merely in respect to unobservable 'interphenomena', the two main classes being 'exhaustive' and 'restrictive' interpretations.

The 'exhaustive' interpretations include the 'interphenomena' while the 'restrictive' interpretations do not. The 'exhaustive' interpretations are represented by interpretations preserving the classical picture of wave or particle while the 'restrictive' interpretations are represented above all by the Bohr-Heisenberg interpretation. The 'exhaustive' interpretations all lead to 'causal anomalies' while the 'restrictive' interpretations do not – in so far the latter are preferable. But even in the latter the 'causal anomalies' are not really overcome but merely incapable of being formulated. [So far the story as told by Reichenbach].

This whole discussion rests on the mistaken view that the mathematical formalism of quantum mechanics *admits* the 'exhaustive' interpretations without leading to inconsistencies or contradictions with the experimental facts. Yet (apart from isomorphisms) a given mathematical formalism admits at the most *one* physical interpretation that is both consistent with the formalism and compatible with the experimental facts concerned. The 'exhaustive' interpretations envisaged by Reichenbach are logically impossible because inconsistent with the formalism [and hence not deserving the name 'interpretations']. On the other hand it is quite true that there exist consistent non-standard interpretations, but they are not equivalent to the standard one and represent a different theory. This applies in particular to the so-called 'causal interpretation' advanced by D. BOHM[11], as shown by TAKABAYASI[12] and the present author[13]. Hence, Reichenbach's explanations appear to me as a vain attempt to find a neutral position that would allow to justify all physical interpretations then under discussion.

In spite of this negative verdict it should be emphasized that in some questions involved in the complicated process of finding the correct physical interpretation Reichenbach took a right stand from the very beginning. Thus, next to Bohr, he was one of the first to refute the idea, advanced particularly by Jordan, that the indeterminacy relations which result from the finite magnitude of the quantum of action are to be explained by the disturbing influence of the measuring devices on the object of measurement: such a disturbing influence exists also in many classical cases.

Finally, it has to be admitted that, notwithstanding the untenability of his theses, Reichenbach still takes a firm stand in defense of materialism. Indeed, Part I of this work ends with these words:

The limitations of scientific interpretations of the world of quanta ... must not be considered as limitations of the power of the human intellect. It is not human ignorance, nor lack of knowledge, which leads to the conditions imposed upon descriptions of the physical world expressed in the laws of quantum mechanics. It is positive knowledge, deep insight into the nature of the atomic world, which constitutes the basis of this strange network of rules ...

Beneath the disguise of a theory of physical knowledge we discern the outlines of a physical world different from what centuries of scientific research had dreamed it to be, but nevertheless demanding recognition as the world of reality.[14]

and on the last page of the work he has this to say about the logical syntax of a physical theory:

Instead of speaking of the structure of the physical world, we may consider the structure of the languages in which this world can be described; such analysis expresses the structure of the world indirectly, but in a more precise way.[15]

This, I suggest, is the essence of a dialectic-materialist conception of logic, much superior to both the normative conception and the vulgar-materialist conception (according to which the laws of logic are the most general laws of Nature).

VI

His most important achievement, in Reichenbach's own view, was his *Theory of Probability* and its offshoot, his *theory of induction*.

Reichenbach's theory of probability originated from a critical examination of existing theories, particularly that of von Kries, and an analysis of the probability concept as used in real science. In accordance with this, his doctor thesis already advocates the *objective* interpretation of 'probability'. In his final theory, probability appears in the simplest case as a numerical relation between two properties, in the most general case as a numerical relation between two coordinated sequences of events. In the simplest case this gives a logico-mathematical calculus combining arithmetic and the logic of classes or predicates. The applicability of the calculus is guaranteed by the fact that all axioms are identically fulfilled by the frequency interpretation.

In contrast to the [then dominant] theory of von Mises [and previous theories] an explicit definition of 'probability' is renounced [as being both impossible and unnecessary] and replaced by implicit (axiomatic) definition. The same axiomatic procedure is also used by Kolmogorov. However, while Kolmogorov's concept of probability is purely mathematical and in fact identical with that of an additive set function, Reichenbach's axiomatic system admits of non-mathematical realizations. Furthermore, Kolmogorov's probability is an absolute probability [one-place function] while Reichenbach's is a relative probability [two-place function], containing the former as the special case that the reference class is the universal class [or rather: that the reference class is always the same and hence identifiable with the universal class]. Apart from this, only Reichenbach's theory (in its general form) allows the development of a theory of the order of probabilistic sequences ('Ordnungstheorie der Wahrscheinlichkeitsfolgen') which appears to be useful for certain applications.

The realistic (antiformalist) attitude of Reichenbach in these questions is expressed very clearly in the following remarks:

Mir scheint der Hauptfehler der Kolmogoroffschen Arbeiten darin zu bestehen, daß bei ihnen die Beziehung zur Anwendung ganz undurchsichtig bleibt. In dieser Beziehung sind die Mathematiker leider mit Blindheit geschlagen ... Auch Kolmogoroff hat keine Ahnung vom Anwendungsproblem, wie schon aus einigen Bemerkungen hervorgeht, in denen er sich auf von Mises' Bemerkungen zum Anwendungsproblem beruft, die sicher der schwächste Teil der ganzen Misesschen Untersuchungen sind. Man kann natürlich irgendeinen mengentheoretischen Sport aus der Wahrscheinlichkeitsrechnung machen ...; aber es bleibt gänzlich schleierhaft, was das mit Physik zu tun hat.[16]

Since this was written, Kolmogorov's theory has been generalized – particularly by the Hungarian mathematician A. Rényi – by taking the concept of *relative* probability as primitive concept, as already done by Reichenbach. Also the arguments used today by Soviet mathematicians in defence of the objective interpretation of 'probability' can almost all be found in Reichenbach's writings. The same applies to the solution of other questions of principle, such as the reduction of equiprobabilities, e.g. in the case the roulette, to the mere existence of a continuous probability function.

To be sure, all these theories concern the classical concept of probability [probability of distribution or coexistence of properties]. The quantum mechanical concept of transition probability is not covered by any of these theories. Here physics has outstripped mathematics.

Reichenbach has used his theory of probability to provide a justification of scientific induction. In this he saw a direct refutation of positivism which does not recognize induction as a scientific method. Marxist philosophers should investigate this problem.[17] I have the impression that Reichenbach's approach is not free of positivist influences and neglects or underestimates the role of practice as criterion of truth.

In the framework of this short contribution it has not been possible to give more than a few hints and suggestions. A critical evaluation of the extensive work of Reichenbach, which is full of original ideas, remains a task to be accomplished by philosophers and physicists working in cooperation. We would be ill-advised would we allow the work of this important and progressive natural philosopher to fall into oblivion.

NOTES

* Translated from 'Hans Reichenbach und die Berliner Schule', in *Naturwissenschaft, Tradition, Fortschritt–Beiheft zur Zeitschrift NTM* (ed. by G. Harig and A. Mette), Berlin, 1963.
[1] H. Reichenbach, *Axiomatik der relativistischen Raum-Zeit-Lehre*, Braunschweig 1924.
[2] All data concerning the appointment of Reichenbach as Professor of Philosophy at Berlin University are taken from the *Akten der Friedrich-Wilhelm-Universität zu Berlin*, Littr. H. No. 1 Vol. **44**, pp. 305–340.
[3] H. Reichenbach, *Wahrscheinlichkeitslehre*, Leiden 1935, p. vi.
[4] *Modern Philosophy of Science, Selected Essays by Hans Reichenbach*, (Translated and edited by Maria Reichenbach), London–New York 1959.
[5] H. Reichenbach, *Logos* **20** (1921), 316–378.
[6] H. Reichenbach, letter to the author, dated 30.3.1936.
[7] H. Reichenbach, letter to the author, dated 13.3.1938.
[8] H. Reichenbach, *Philosophie der Raum-Zeit-Lehre*, Berlin–Leipzig 1928, pp. 1–4.
[9] Cf. note 1.
[10] H. Reichenbach, *Philosophic Foundations of Quantum Mechanics*, Berkeley and Los Angeles 1944.
[11] D. Bohm, 'Causal interpretation of quantum mechanics', *Phys. Rev.* **85** (1952), 166, 180.
[12] T. Takabayasi, *Prog. Theor. Phys.* **8** (1952), 143, **9** (1953), 187.

[13] M. Strauss, Discussion Remark at the Jahreshauptversammlung der Physikalischen Gesellschaft der DDR, Dresden 1955.

[14] H. Reichenbach, *loc. cit.* (note 10), 44.

[15] H. Reichenbach, *loc. cit.* (note 10), 177.

[16] H. Reichenbach, letter to the author, dated 5.7.1936.

[17] [Cf. G. I. Ruzavin, 'Die Wahrscheinlichkeitslogik und ihre Rolle in der wissenschaft-lichen Forschung' in *Studien zur Logik der wissenschaftlichen Erkenntnis* (ed. by G. Kroeber), Berlin 1967.]

CHAPTER XXIV

TWO NOTES ON H. REICHENBACH'S LOGIC
OF QUANTUM MECHANICS*

INTRODUCTON AND SUMMARY

In his new book[1] H. Reichenbach advocates the use of three-valued logic for the language of quantum mechanics. It is by this means, he says, that we can best suppress the causal anomalies which are known to arise when this theory is stated within the syntactic framework employed in classical mechanics[2].

In the two notes to follow I propose to show that the use of three-valued logic is neither necessary nor sufficient for the suppression of the causal anomalies.

In the first note Reichenbach's three-valued language is translated into a two-valued language which in fact we can identify with the ordinary language of quantum mechanics, viz., the language most widely used by quantum physicists. From the nature of this translation it will become clear that if a causal anomaly cannot be construed in the original language it cannot be construed in the translation either. Hence it may be concluded that three-valued logic is not necessary for the suppression of causal anomalies.

Incidentally, our translation may be taken as an example of a general method by which three-valued logic can be interpreted in terms of two-valued logic.

In the second note I discuss Reichenbach's application of three-valued logic to a well-known paradox. Reichenbach is correct in saying that this paradox can be reduced to a causal anomaly. However, his way of suppressing this causal anomaly implies the statement of another causal anomaly which is even more serious.

My objection to the use of a three-valued language for quantum mechanics does not imply, and is not based on, a general rejection of three-(or more-) valued languages. Nor should it be mistaken to imply an unfavourable opinion, on my part, on Reichenbach's book. I welcome this book as a most valuable contribution to the logical analysis of quantum mechanics.

I. TRANSLATION OF REICHENBACH'S
THREE-VALUED LANGUAGE INTO THE ORDINARY LANGUAGE
OF QUANTUM MECHANICS

1. *Reichenbach's Three-Valued Language L_3*

Reichenbach uses the letters

$$A, B, C, \ldots$$

as sentential symbols (constants and variables) of his three-valued language L_3. His three truth-values are 'T' (for 'true'), 'I' (for 'indeterminate'[4]), and 'F' (for 'false'). These three predicates are considered to be mutually exclusive and to form a complete alternative for the sentences of L_3.

Sentential connectives are introduced by means of the following truth-tables.

TABLE A

A	Cyclical negation $\sim A$	Diametrical negation $-A$	Complete negation $\bar{A}$
T	I	F	I
I	F	I	T
F	T	T	T

TABLE B

A	B	$A \vee B$	$A \cdot B$	$A \supset B$	$A \rightarrow B$	$A \Rightarrow B$	$A \equiv B$	$A \equiv B$
						T	T	T
T	T	T	T	T	T	T	T	T
T	I	T	I	I	F	I	I	F
T	F	T	F	F	F	F	F	F
I	T	T	I	T	T	I	I	F
I	I	I	I	T	T	I	T	T
I	F	I	F	I	T	I	I	F
F	T	T	F	T	T	I	F	F
F	I	I	F	T	T	I	I	F
F	F	F	F	T	T	I	T	T

The rules of inference consist in the application of the modus ponens to each of the three implications[5], i.e.

$$\frac{A \overset{A}{\supset} B}{B} \quad \frac{A \overset{A}{\rightarrow} B}{B} \quad \frac{A \overset{A}{\Rightarrow} B}{B}.$$
(R1)

The assertion that 'A' is indeterminate can be expressed in the object-language by

$$\sim \sim A.$$
(R2)

The relationship of complementarity in quantum mechanics is expressed by the formula

$$A \vee \sim A \rightarrow \sim \sim B$$
(R3)

which is easily proved to be equivalent to

$$B \vee \sim B \rightarrow \sim \sim A.$$

2. *Translation of L_3 into L_2*

In the following the letters

$$a, b, c, ...; \quad \alpha, \beta, \gamma, ...$$

are sentential symbols of a two-valued language L_2. The signs between these symbols are the ordinary sentential connectives.

For L_2 we assume the postulates

$$\begin{aligned} &\alpha \supset a \\ &\beta \supset b \\ &\gamma \supset c \end{aligned}$$
(2.1)

. . . .

We now translate L_3 into L_2 as follows. First, we translate

$$\begin{aligned} A \quad &\text{into} \quad a \cdot \alpha \\ B \quad &\text{into} \quad b \cdot \beta \\ C \quad &\text{into} \quad c \cdot \gamma \end{aligned}$$
(2.2)

. . . .

The following Table shows the co-ordination of truth-values in the case of the negations of L_3.

a	α	A	$\sim A$	$-A$	$\bar{A}$	$\sim\sim A$	$--A$	$\bar{\bar{A}}$
+	+	T	I	F	I	F	T	T
−	−	I	F	I	T	T	I	I
+	−	F	T	T	T	I	F	I

(2.3)

Here, '+' and '−' stand for the terms 'true' and 'false' of two-valued logic. The line with $(a, \alpha)=(-, +)$ has been omitted since this case is excluded by the postulates (2.1).

We now translate the negations of L_3 in such a way that a true sentence of L_3 is translated into a true sentence of L_2, and vice versa. It follows that if a sentence of L_3 happens to be indeterminate its translation in L_2 is a false sentence. From the Table (2.3) we thus obtain the following translations.

$$\sim A \quad \text{into} \quad a \cdot \sim \alpha \qquad\qquad\qquad\qquad\qquad (2.3a)$$
$$- A \quad \text{into} \quad a \cdot \sim \alpha \qquad\qquad\qquad\qquad\qquad (2.3b)$$
$$\bar{A} \quad \text{into} \quad (a \cdot \sim \alpha) \vee (\sim a \cdot \sim \alpha) \quad \therefore \sim a \quad (2.3c)$$
$$\sim\sim A \quad \text{into} \quad \sim a \cdot \sim \alpha \qquad\qquad\qquad \therefore \sim a \quad (2.3d)$$
$$-- A \quad \text{into} \quad a \cdot \alpha \qquad\qquad\qquad\qquad \therefore \alpha \quad (2.3e)$$
$$\bar{\bar{A}} \quad \text{into} \quad a \cdot \alpha \qquad\qquad\qquad\qquad \therefore \alpha \quad (2.3f)$$

The application of the same principle to the other sentential connectives of L_3 leads to the following translations.

$$A \vee B \quad \text{into} \quad (a \cdot \alpha) \vee (b \cdot \beta) \qquad\qquad (2.4)$$
$$A \cdot B \quad \text{into} \quad (a \cdot \alpha) \cdot (b \cdot \beta) \qquad\qquad (2.5)$$
$$A \supset B \quad \text{into} \quad (a \vee b) \cdot (a \cdot \alpha \supset \beta) \quad (2.6a)$$
$$A \rightarrow B \quad \text{into} \quad a \cdot \alpha \supset b \cdot \beta \qquad\qquad (2.6b)$$
$$A \Rightarrow B \quad \text{into} \quad (a \cdot \alpha) \cdot (b \cdot \beta) \qquad\qquad (2.6c)$$
$$A \equiv B \quad \text{into} \quad (a \equiv b) \cdot (\alpha \equiv \beta) \qquad (2.7a)$$
$$A \equiv B \quad \text{into} \quad (a \equiv b) \cdot (\alpha \equiv \beta) \qquad (2.7b)$$

As is to be expected, our translation gives a many-one correlation between L_3 and L_2, not a one-one correlation. The translation, however, preserves the relationship of consequence. Moreover, the relationship of non-consequence is also preserved in the translation from L_3 into L_2. Indeed, the false and indeterminate sentences of L_3 cannot be used as premises of the modus ponens in L_3, and these sentences are correlated to false sentences in L_2 where the latter cannot be used as premises either. Hence if a conclusion leading to a causal anomaly is suppressed in L_3 by the use of indeterminate sentences this conclusion cannot be drawn in L_2 either.

From (2.6b), (2.3d), (2.3a), and (2.1) we obtain the translation of the formula (R3) expressing the relationship of complementarity in quantum mechanics, namely

$$a \supset \sim b. \tag{2.8}$$

3. L_2 Identified with the Ordinary Language of Quantum Mechanics

For L_2 we assume the following rules of interpretation.

(3.1) 'a' is short for 'At the instant t the system s is in a state where the entity A has a definite value'

or

'$(\exists \alpha) P^A(s, t, \alpha)$',

(3.2) 'α' is short for 'At the instant t the system s is in a state where the entity A has the value α'

or

'$P^A(s, t, \alpha)$'.

From this interpretation the postulates (2.1) follow.

Reichenbach's formula (R2) expressing the case of indeterminacy has been translated into '$\sim a$', i.e., 'At the instant t the system is not in a state where the entity A has a definite value'.

The translation (2.8) of Reichenbach's formula (R3) expressing the relationship of complementarity in quantum mechanics now reads 'If at the instant t the system s is in a state where the entity A has a definite value then at the instant t the system s is not in a state where the entity B has a definite value'.

Thus, we have identified L_2 with the ordinary language of quantum mechanics.

4. The Syntactic Framework of Classical Mechanics

In classical mechanics the sentences 'a', 'b', ... are always true. This allows us to consider the physical entities as functions of time and to write

'$A(s, t) = \alpha$'   instead of   '$P^A(s, t, \alpha)$'.

The negation of '$a \cdot \alpha$', i.e. '$a \cdot \sim \alpha \vee \sim a$', then reduces to the first term of this disjunction, i.e. '$a \cdot \sim \alpha$' for which we usually write

$$\text{'}A(s, t) \neq \alpha \text{'} \quad \text{meaning} \quad \text{'}A(s, t) = \sigma \quad \text{and} \quad \sigma \neq \alpha \text{'}.$$

5. *Reichenbach's Three-Valued Logic as a Substitute for the Transition from Functors to Predicates'*

Reichenbach's sentential symbol 'A' can be considered as short for '$A(s, t) = \alpha$' which may be read 'At the instant t the entity A of s has the value α'. The negation '$A(s, t) \neq \alpha$' is considered to be the proper negation. This leaves no possibility to express the case of indeterminacy if two-valued logic is employed. Thus Reichenbach's use of three-valued logic may be considered as a substitute for the use of predicates instead of functors.

6. *Reichenbach's Own Reduction of his Three-Valued Language to a Two-Valued Language*[6]

This reduction is given by the table

Observational language		Quantum mechanical language
m_u	u	U
T	T	T
T	F	F
F	T	I
F	F	I

The following interpretation is given.

'm_u'   is short for 'A measurement of u is made'
'u'   is short for 'The measuring instrument indicates the value u'
'U'   is short for 'The value of the entity immediately after the measurement is u'.

This interpretation, apart from being incomplete by the omission of a reference to the system subjected to measurement, is too narrow as far as 'U' is concerned. If we test a prognosis it is the value at the beginning of the measurement we are interested in. Moreover, if the reference to the system subjected to measurement is not omitted we obtain the reduction sentence

$$m_u \cdot U \equiv u \tag{6.1}$$

which contradicts the third line of the table. Indeed, 'm_u' is now short for 'A measurement of the entity U of the system s is made', and 'u' is short for 'The measuring instrument which is used to measure U of s indicates the value u'. Obviously, if 'm_u' is false 'u' cannot be true. In fact, 'u' implies 'm_u'.

The reduction of the quantum mechanical language to the observational language, correctly given by (6.1), leaves the truth-value of 'U' undetermined if no measurement is made. However, this does not mean that the truth-value of 'U' cannot be inferrred from the accepted truth-values of other sentences. For instance, if the entity V is complementary to the entity U, and the sentence 'V' is true, 'U' must be considered false because otherwise contradictions can be deduced.

In conclusion, I wish to emphasize that the translation of Reichenbach's three-valued language given in §2 and §3 has nothing to do with the question of its reduction to an observational language.

II. THREE-VALUED LOGIC AND THE SUPPRESSION OF CAUSAL ANOMALIES

1. *The Experimental Arrangement*

With REICHENBACH[7] let us consider the following experimental arrangement[8].

A diaphragm with two narrow parallel slits B_1 and B_2 is placed between a scintillation screen C and a source A of electrons or other atomic particles. For the sake of simplicity let us assume that diaphragm and screen form two concentric spheres with the source at their common centre. The screen may be assumed to be transparent so that the flashes caused by the particles hitting the screen can be observed from outside. The rate of emission (the intensity of the beam) is assumed to be constant and so low that there is no interaction between the particles of the beam. The slits can be closed separately by means of flaps.

Exp. 1. Slit B_1 is open, slit B_2 is closed. The flashes on the screen are observed for, say, ten minutes, and their positions marked on the screen. Their distribution shows the diffraction pattern predicted by the theory and may hence be represented by the predicted distribution function

$$d_1(S), \quad \int d_1(S)\, dS = N_1 \qquad\qquad (1.1)$$

where $S=(x, y)$ is a point on the screen, and N_1 the total number of flashes observed.

Exp. 2. Slit B_1 is closed, slit B_2 is open. The flashes are again observed for ten minutes. The distribution function in this case is

$$d_2(S), \quad \int d_2(S)\, dS = N_2. \qquad\qquad (1.2)$$

Exp. 3. Both slits are open. The flashes are again observed for ten minutes. The distribution function in this case be

$$d_3(S), \quad \int d_3(S)\, dS = N_3, \qquad\qquad (1.3)$$

N_3 being the total number of flashes observed in this case.

2. *Relations Between the Experimental Results*

$d_3(S)$ shows an interference pattern. For later reference we note the following relations between d_3, d_1 and d_2.

$$N_3 = N_1 + N_2 \qquad\qquad (2.1)$$
$$d_3(S) \neq d_1(S) + d_2(S). \qquad\qquad (2.2)$$

As a consequence of (2.1) and (2.2) we have

$$\text{for some } S: d_3(S) > d_1(S) + d_2(S). \qquad\qquad (2.3)$$

3. *The Paradox*

Since in exp. 3. a particle can pass through either slit it is concluded that the distribution function d_3 should be the sum of d_1 and d_2, i.e.

$$d_3(S) = d_1(S) + d_2(S). \qquad\qquad (3.1)$$

This contradicts (2.2). Thus (3.1) and (2.2) form a paradox.

4. Re-statement of the Paradox in Terms of Probability

$d_1(S)$ is a direct measure of the probability density that a particle emitted from A passes through the slit B_1 and hits the screen at S;

$$d_1(S) = N_0 \operatorname{prob}(A; B_1 \cdot S). \tag{4.1}$$

Similarly

$$d_2(S) = N_0 \operatorname{prob}(A; B_2 \cdot S), \tag{4.2}$$
$$d_3(S) = N_0 \operatorname{prob}(A; [B_1 \vee B_2] \cdot S). \tag{4.3}$$

N_0 is a constant which depends on the intensity of the beam and the duration of the observation.

Resolving the square bracket and applying the general theorem of addition we obtain

$$d_3(S) = d_1(S) + d_2(S) - N_0 \operatorname{prob}(A; B_1 \cdot B_2 \cdot S). \tag{4.4}$$

The last term represent the particles which pass through both slits. We assume it to be zero,

$$\operatorname{prob}(A; B_1 \cdot B_2 \cdot S) = 0 \tag{4.5}$$

and obtain again

$$d_3(S) = d_1(S) + d_2(S). \tag{4.6}$$

5. The Wave-Properties Do Not Solve the Paradox

It may be argued that the paradox is due to the corpuscular language employed and the neglect of the wave-properties of our atomic particles. It is true that the distribution functions d_1, d_2, and d_3 can be deduced from the assumption that the particles behave like waves when passing through the slits. However, these functions represent the distribution of individual flashes on the screen the occurrence of which cannot be explained in the wave language, at least not without introducing causal anomalies. We are therefore justified in using the corpuscular language.

The corpuscular behaviour on hitting the screen, however, does not imply a corpuscular behaviour in the passing through the slits. It may therefore be concluded that an atomic particle may pass through both slits simultaneously and that our assumption (4.5) is wrong. This consideration is not a way out.

Indeed, whether the probability (4.5) is zero or not, it follows from (4.4) that

$$\text{for no } S: d_3(S) > d_1(S) + d_2(S) \tag{5.1}$$

which is contradicted by (2.3).

6. *Possible Ways of Preventing the Paradox*

From the formal point of view the following ways are open to us to prevent the conclusion (4.6) from being established.

(i) The distribution functions d_1, d_2, d_3, or some of them, are incorrectly represented by the probability expressions of § 4.

(ii) For special reasons we are not allowed to apply the addition theorem to the probability expression in (4.3).

Reichenbach does see the first way, but he actually chooses the second way.

The first way divides into two routes:

(ia) The first-place arguments in the probability expressions of § 4 are incomplete;

(ib) '$B_1 \vee B_2$' in (4.3) is an incorrect representation of the condition that both slits are open.

It is the first route which is seen, but not followed up, by Reichenbach. It leads to the reduction of the paradox to a causal anomaly.

7. *Reduction of the Paradox to a Causal Anomaly*

In the probability expressions of § 4 the first-place arguments, representing the experimental conditions, are all the same. Yet these conditions are actually different, different slits being open or closed.

The correct probability expressions for the distribution functions are

$$d_1(S) = N_0 \operatorname{prob}(A \cdot B_1^{\mathrm{op}} \cdot B_2^{\mathrm{cl}}; B_1 \cdot S) \tag{7.1}$$
$$d_2(S) = N_0 \operatorname{prob}(A \cdot B_1^{\mathrm{cl}} \cdot B_2^{\mathrm{op}}; B_2 \cdot S) \tag{7.2}$$
$$d_3(S) = N_0 \operatorname{prob}(A \cdot B_1^{\mathrm{op}} \cdot B_2^{\mathrm{op}}; [B_1 \vee B_2] \cdot S) \tag{7.3}$$

If we now apply the addition theorem to (7.3) and compare the result with the experimental result (2.2), we merely obtain

$$\operatorname{prob}(A \cdot B_1^{\mathrm{op}} \cdot B_2^{\mathrm{op}}; B_1 \cdot S) + \operatorname{prob}(A \cdot B_1^{\mathrm{op}} \cdot B_2^{\mathrm{op}}; B_2 \cdot S) \neq$$
$$\operatorname{prob}(A \cdot B_1^{\mathrm{op}} \cdot B_2^{\mathrm{cl}}; B_1 \cdot S) + \operatorname{prob}(A \cdot B_1^{\mathrm{cl}} \cdot B_2^{\mathrm{op}}; B_2 \cdot S) \tag{7.4}$$

and hence, for reasons of symmetry,

$$\text{prob}(A \cdot B_1^{\text{op}} \cdot B_2^{\text{op}}; B_1 \cdot S) \neq \text{prob}(A \cdot B_1^{\text{op}} \cdot B_2^{\text{cl}}; B_1 \cdot S). \qquad (7.5)$$

This means that the probability that a particle passes through slit B_1 and reaches the screen at S, depends on the slit B_2 being open or closed. This constitutes 'action at a distance', and hence a causal anomaly.

8. *Reichenbach's Solution*

Reichenbach suppresses the causal anomaly (7.5) by forbidding us to apply the addition theorem to (7.3), on the ground that '$B_1 \vee B_2$' is indeterminate[9] since no observation has taken place at the slits.

To this way of arguing it might be objected that only 'B_1' and 'B_2' are indeterminate, but not '$B_1 \vee B_2$'. This, however, is impossible, according to Reichenbach's definition of the three-valued disjunction.

Yet this solution has a serious consequence. If '$B_1 \vee B_2$' is indeterminate then the sentence

$$S \rightarrow B_1 \vee B_2 \qquad (8.1)$$

is false for all particles for which 'S' is true, i.e., all particles which have hit the screen; if we use any of the other implication signs, the implication becomes indeterminate. Hence, in Reichenbach's language it is *not possible to infer from the observed flash on the screen that the particle causing the flash has passed through the slits in the diaphragm.* This, in my opinion, is a causal anomaly even worse than the one suppressed.

Reichenbach's solution leads to another query. If '$B_1 \vee B_2$' is indeterminate, the probability expression for d_3 should be written

$$d_3(S) = N_0 \, \text{prob}(A \cdot B_1^{\text{op}} \cdot B_2^{\text{op}}; \sim \sim [B_1 \vee B_2] \cdot S).$$

It should be possible to transform this expression into simpler ones. However, Reichenbach has not so far developed a probability calculus applicable to a three-valued language. Only if this is done can the implications of his proposed solution be fully discussed.

NOTES

* Published for the first time. The paper was written in 1945.
[1] H. Reichenbach, *Philosophic Foundations of Quantum Mechanics*, University of California Press, 1944.
[2] Its relevant features will be stated in § 4.
[3] Cf. *loc. cit.*, § 32.
[4] *Reichenbach's* 'indeterminate' is conceived as a semantic term; it must not be confused with the syntactic term 'indeterminate' as defined by Carnap.
[5] Cf. *loc. cit.*, p. 152.
[6] Cf. *loc. cit.*, § 30.
[7] Cf. *loc. cit.*, § 7.
[8] This arrangement has often been used to illustrate the peculiarities of quantum mechanics.
[9] Reichenbach writes the probability expressions in such a way that the disjunction appears in the first-place argument. This makes no difference to the questions discussed.

SYNTHESE LIBRARY

Monographs on Epistemology, Logic, Methodology,
Philosophy of Science, Sociology of Science and of Knowledge, and on the
Mathematical Methods of Social and Behavioral Sciences

Editors:

DONALD DAVIDSON (Rockefeller University and Princeton University)
JAAKKO HINTIKKA (Academy of Finland and Stanford University)
GABRIËL NUCHELMANS (University of Leyden)
WESLEY C. SALMON (Indiana University)

‡DONALD DAVIDSON and GILBERT HARMAN (eds.), *Semantics of Natural Language.* 1972,
X + 769 pp. Dfl. 110,—
‡ROGER C. BUCK and ROBERT S. COHEN (eds.), *Boston Studies in the Philosophy of
Science.* Volume VIII: *PSA 1970. In Memory of Rudolf Carnap.* 1971, LXVI + 617 pp.
Dfl. 120,—
‡STEPHEN TOULMIN and HARRY WOOLF (eds.), *Norwood Russell Hanson: What I Do Not
Believe, and Other Essays.* 1971, XII + 390 pp. Dfl. 90,—
‡YEHOSHUA BAR-HILLEL (ed.), *Pragmatics of Natural Languages.* 1971, VII + 231 pp.
Dfl. 50,—
‡ROBERT S. COHEN and MARX W. WARTOFSKY (eds.), *Boston Studies in the Philosophy of
Science.* Vol. VII: Milič Čapek: *Bergson and Modern Physics.* 1971, XV + 414 pp.
Dfl. 70,—
‡CARL R. KORDIG, *The Justification of Scientific Change.* 1971, XIV + 119 pp. Dfl. 33,—
‡JOSEPH D. SNEED, *The Logical Structure of Mathematical Physics.* 1971, XV + 311 pp.
Dfl. 70,—
‡JEAN-LOUIS KRIVINE, *Introduction to Axiomatic Set Theory.* 1971, VII + 98 pp. Dfl. 28,—
‡RISTO HILPINEN (ed.), *Deontic Logic: Introductory and Systematic Readings.* 1971,
VII + 182 pp. Dfl. 45,—
‡EVERT W. BETH, *Aspects of Modern Logic.* 1970, XI + 176 pp. Dfl. 42,—
‡PAUL WEINGARTNER and GERHARD ZECHA (eds.), *Induction, Physics, and Ethics. Pro-
ceedings and Discussions of the 1968 Salzburg Colloquium in the Philosophy of Science.*
1970, X + 382 pp. Dfl. 65,—
‡ROLF A. EBERLE, *Nominalistic Systems.* 1970, IX + 217 pp. Dfl. 42,—
‡JAAKKO HINTIKKA and PATRICK SUPPES, *Information and Inference.* 1970, X + 336 pp.
Dfl. 60,—
‡KAREL LAMBERT, *Philosophical Problems in Logic. Some Recent Developments.* 1970,
VII + 176 pp. Dfl. 38,—
‡P. V. TAVANEC (ed.), *Problems of the Logic of Scientific Knowledge.* 1969, XII + 429 pp.
Dfl. 95,—
‡ROBERT S. COHEN and RAYMOND J. SEEGER (eds.), *Boston Studies in the Philosophy of
Science.* Volume VI: *Ernst Mach: Physicist and Philosopher.* 1970, VIII + 295 pp.
Dfl. 38,—

‡MARSHALL SWAIN (ed.), *Induction, Acceptance, and Rational Belief.* 1970, VII + 232 pp.
Dfl. 40,—

‡NICHOLAS RESCHER *et al.* (eds.), *Essays in Honor of Carl G. Hempel. A Tribute on the Occasion of his Sixty-Fifth Birthday.* 1969, VII + 272 pp.
Dfl. 50,—

‡PATRICK SUPPES, *Studies in the Methodology and Foundations of Science. Selected Papers from 1951 to 1969.* 1969, XII + 473 pp.
Dfl. 72,—

‡JAAKKO HINTIKKA, *Models for Modalities. Selected Essays.* 1969, IX + 220 pp. Dfl. 34,—

‡D. DAVIDSON and J. HINTIKKA (eds.), *Words and Objections: Essays on the Work of W. V. Quine.* 1969, VIII + 366 pp.
Dfl. 48,—

‡J. W. DAVIS, D. J. HOCKNEY and W. K. WILSON (eds.), *Philosophical Logic.* 1969, VIII + 277 pp.
Dfl. 45,—

‡ROBERT S. COHEN and MARX W. WARTOFSKY (eds.), *Boston Studies in the Philosophy of Science.* Volume V: *Proceedings of the Boston Colloquium for the Philosophy of Science 1966/1968.* 1969, VIII + 482 pp.
Dfl. 60,—

‡ROBERT S. COHEN and MARX W. WARTOFSKY (eds.), *Boston Studies in the Philosophy of Science.* Volume IV: *Proceedings of the Boston Colloquium for the Philosophy of Science 1966/1968.* 1969, VIII + 537 pp.
Dfl. 72,—

‡NICHOLAS RESCHER, *Topics in Philosophical Logic.* 1968, XIV + 347 pp. Dfl. 70,—

‡GÜNTHER PATZIG, *Aristotle's Theory of the Syllogism. A Logical-Philological Study of Book A of the Prior Analytics.* 1968, XVII + 215 pp.
Dfl. 48,—

‡C. D. BROAD, *Induction, Probability, and Causation. Selected Papers.* 1968, XI + 296 pp.
Dfl. 54,—

‡ROBERT S. COHEN and MARX W. WARTOFSKY (eds.), *Boston Studies in the Philosophy of Science.* Volume III: *Proceedings of the Boston Colloquium for the Philosophy of Science 1964/1966.* 1967, XLIX + 489 pp.
Dfl. 70,—

‡GUIDO KÜNG, *Ontology and the Logistic Analysis of Language. An Enquiry into the Contemporary Views on Universals.* 1967, XI + 210 pp.
Dfl. 41,—

*EVERT W. BETH and JEAN PIAGET, *Mathematical Epistemology and Psychology.* 1966, XXII + 326 pp.
Dfl. 63,—

*EVERT W. BETH, *Mathematical Thought. An Introduction to the Philosophy of Mathematics.* 1965, XII + 208 pp.
Dfl. 37,—

‡PAUL LORENZEN, *Formal Logic.* 1965, VIII + 123 pp.
Dfl. 26,—

‡GEORGES GURVITCH, *The Spectrum of Social Time.* 1964, XXVI + 152 pp. Dfl. 25,—

‡A. A. ZINOV'EV, *Philosophical Problems of Many-Valued Logic.* 1963, XIV + 155 pp.
Dfl. 32,—

‡MARX W. WARTOFSKY (ed.), *Boston Studies in the Philosophy of Science.* Volume I: *Proceedings of the Boston Colloquium for the Philosophy of Science, 1961–1962.* 1963, VIII + 212 pp.
Dfl. 26,50

‡B. H. KAZEMIER and D. VUYSJE (eds.), *Logic and Language. Studies dedicated to Professor Rudolf Carnap on the Occasion of his Seventieth Birthday.* 1962, VI + 256 pp. Dfl. 35,—

*EVERT W. BETH, *Formal Methods. An Introduction to Symbolic Logic and to the Study of Effective Operations in Arithmetic and Logic.* 1962, XIV + 170 pp.
Dfl. 35,—

*HANS FREUDENTHAL (ed.), *The Concept and the Role of the Model in Mathematics and Natural and Social Sciences. Proceedings of a Colloquium held at Utrecht, The Netherlands, January 1960.* 1961, VI + 194 pp.
Dfl. 34,—

‡P. L. R. GUIRAUD, *Problèmes et méthodes de la statistique linguistique.* 1960, VI + 146 pp.
Dfl. 28,—

*J. M. BOCHEŃSKI, *A Precis of Mathematical Logic.* 1959, X + 100 pp.
Dfl. 23,—

SYNTHESE HISTORICAL LIBRARY

Text and Studies
in the History of Logic and Philosophy

Editors:

N. KRETZMANN (Cornell University)
G. NUCHELMANS (University of Leyden)
L. M. DE RIJK (University of Leyden)

‡LEWIS WHITE BECK (ed.), *Proceedings of the Third International Kant Congress, held at the University of Rochester, March 30–April 4, 1970.* 1972, XI + 718 pp.　　Dfl. 160,—
‡KARL WOLF and PAUL WEINGARTNER (eds.), *Ernst Mally: Logische Schriften.* 1971, X + 340 pp.　　　　　　　　　　　　　　　　　　　　　Dfl. 80,—
‡LEROY E. LOEMKER (ed.), *Gottfried Wilhelm Leibnitz: Philosophical Papers and Letters.* A Selection Translated and Edited, with an Introduction. 1969, XII + 736 pp.
　　　　　　　　　　　　　　　　　　　　　　　　　　　　Dfl. 125,—
‡M. T. BEONIO-BROCCHIERI FUMAGALLI, *The Logic of Abelard.* Translated from the Italian. 1969, IX + 101 pp.　　　　　　　　　　　　　　　Dfl. 27,—

Sole Distributors in the U.S.A. and Canada:
*GORDON & BREACH, INC., 440 Park Avenue South, New York, N.Y. 10016
‡HUMANITIES PRESS, INC., 303 Park Avenue South, New York, N.Y. 10010